HOW TO BUILD THE PERFECT BACKYARD HOMESTEAD

HOW TO BUILD THE PERFECT BACKYARD HOMESTEAD

A COMPREHENSIVE GUIDE TO SELF-SUFFICIENCY WITH GARDENING, RAISING CHICKENS, BEEKEEPING, FOOD PRESERVATION, BASIC CARPENTRY, AND MORE

SELF-SUFFICIENT LIVING

BOOK 1

ANTHONY BENNETT

B

Book Bound
STUDIOS

To all who dream of a greener, more self-sufficient life—this book is for you. It's a tribute to the spirit of resilience, to families seeking harmony with nature, and to individuals daring to embrace a sustainable lifestyle. May this book guide you from the seed of an idea to the flourishing reality of your own backyard oasis. With heartfelt thanks to my family for their support and to the earth that sustains us all.

In the spring, at the end of the day, you should smell like dirt.

— MARGARET ATWOOD

CONTENTS

INTRODUCTION TO BACKYARD HOMESTEADING

A lush garden with a wooden shed.

The Philosophy of Homesteading

Embarking on the journey of backyard homesteading is not just about transforming your outdoor space; it's about embracing a philosophy that intertwines self-sufficiency, sustainability, and a deep connection with the natural world. This philosophy is rooted in the belief that even the smallest plot of land can be a powerful source of food, joy, and life. It's about seeing potential where others see limitation, about nurturing the earth and, in return, being nurtured by it.

At the heart of homesteading is the desire to reduce one's carbon footprint and live less dependent on the systems that, while convenient, often distance us from the natural cycles of life. It's about learning to work with these cycles, understand the seasons' rhythm, and respect the resources we so often take for granted. This approach to living doesn't just benefit the individual or the immediate family; it extends its positive impact on the community and the environment.

The philosophy of homesteading encourages a mindset of growth, resilience, and creativity. It's about problem-solving, whether finding ways to maximize a small space for vegetable growing, devising water-saving techniques, or repurposing materials for garden structures. It teaches patience and persistence, as not every endeavor will be successful on the first try, but every failure is a lesson learned and an opportunity for growth.

Moreover, homesteading is about community. It fosters a sense of connection to the land and those around us. Sharing harvests, exchanging seeds, and passing on knowledge are all integral aspects of the homesteading philosophy. It's a way of living that encourages cooperation over competition, creating a network of support that benefits everyone involved.

In embracing the homesteading philosophy, one also embraces a learning life. There is always a new skill to master, a new plant to grow, or a new sustainability practice to implement. This continuous journey of discovery keeps the homesteader engaged and connected to their environment meaningfully.

As we delve deeper into the practical aspects of what you can achieve in your backyard, remember that this philosophy is the foundation of all these endeavors. It's not just about the tangible outcomes but the values, principles, and connections these activities embody. Whether you're planting your first vegetable garden, setting up a rainwater collection system, or building a chicken coop, you're participating in a way of life that is both ancient and increasingly relevant in our modern world.

Backyard homesteading opens up a world of possibilities right at your doorstep. This section aims to illuminate your backyard's vast potential, transforming it from a mere outdoor space to a thriving homestead. Whether you have a sprawling lawn or a modest patch of green, the essence of homesteading is about making the most of what you have. Let's explore what you can achieve with dedication, creativity, and elbow grease.

One of the most rewarding aspects of backyard homesteading is the ability to grow your food. This is wider than sprawling vegetable gardens, although they are a fantastic start. To maximize space, you can cultivate various fruits and vegetables in raised beds, containers, or even vertical gardens. Herbs, too, can flourish in small pots on windowsills, adding fresh flavors to your meals. With some planning, you can enjoy fresh produce, reduce your grocery bill, and increase your self-sufficiency.

While the idea of livestock might conjure images of vast farmlands, certain animals are well-suited for smaller spaces. Chickens, for instance, can provide a steady supply of fresh eggs while contributing to pest control and composting. Rabbits and bees are other livestock that can be accommodated in a backyard, offering meat, wool, and honey. It's important to check local regulations and consider your capacity to care for these animals before diving in.

Homesteading is not just about food production; it's about creating a sustainable ecosystem in your backyard. Composting kitchen scraps and yard waste can produce rich soil for your garden, reducing waste and the need for chemical fertilizers. Rainwater harvesting systems can supplement your water supply, conserving this precious resource. Even incorporating native plants and creating habitats can attract beneficial wildlife, promote biodiversity, and aid in pest control.

For the more ambitious homesteader, exploring renewable energy options can reduce reliance on external power sources. Solar panels, for instance, can be a viable option for generating electricity or heating water. While the initial investment might be significant, the long-term benefits and potential savings are considerable. Though less common, wind turbines can also be an option depending on your location and property size.

Backyard homesteading also opens up opportunities for crafting and DIY projects. There are many projects to embark on, from building chicken coops and raised garden beds to creating homemade soaps and preserves. These activities provide practical benefits and allow for creative expression and the satisfaction of building something with your hands. Regardless of size, your backyard can be transformed into a productive and sustainable homestead. It's about leveraging your resources to create a self-sufficient space that meets your needs and reflects your values. As we move forward, planning will be crucial to effectively utilize your space and resources, turning your homesteading dreams into reality.

Planning Your Homestead Layout

Backyard homesteading is an exciting venture that promises a bounty of fresh produce and possibly livestock and a fulfilling connection to the earth and the food we consume. However, before diving into the practical aspects of planting and animal care, meticulously planning your homestead layout is crucial for laying a solid foundation. This step is vital for maximizing efficiency, ensuring sustainability, and, ultimately, enjoying the fruits of your labor.

First and foremost, assess the space available in your backyard. Every square foot counts, and understanding the dimensions of your land will help you make informed decisions about what you can realistically achieve. When evaluating your space, consider factors such as sunlight exposure, soil quality, and access to water. These elements are critical in determining which areas are best suited for gardening, which spots are ideal for composting, and where you could house small livestock.

Next, think about the kind of homestead you envision. Are you leaning more toward vegetable gardening, or are you interested in raising chickens for eggs? You may combine both with other ventures like beekeeping or growing medicinal herbs. Your interests and goals will significantly influence your layout plan. For instance, a vegetable garden requires well-draining soil and at least six hours of direct sunlight daily, while chickens need a secure coop and a fenced area to roam.

Once you have a clear idea of what you want to include in your homestead, it's time to sketch a layout. This doesn't have to be a professional blueprint; a simple drawing that outlines where each element will go is sufficient. Make sure to allocate space for pathways and consider the ease of access to each area.

Efficient movement around your homestead will save you time and energy in the long run.

Incorporating companion planting and crop rotation into your layout can significantly benefit your garden's health and yield. Companion planting involves placing plants together that can help each other grow, deter pests, or enhance flavor. Crop rotation, on the other hand, prevents soil depletion and reduces the risk of disease. Planning these strategies from the outset can lead to a more resilient and productive homestead.

Lastly, remember that flexibility is critical. Your initial layout might need adjustments as you learn and grow in your homesteading journey. You may discover that a particular crop thrives in your soil, prompting you to allocate more space next season. Or you'll expand your chicken coop as you become more experienced in poultry care. Embrace these changes as they are a natural part of homesteading.

In conclusion, planning your homestead layout is a crucial step that sets the stage for a successful and rewarding backyard homesteading experience. By assessing your space, defining your goals, and sketching a thoughtful layout, you'll be well on your way to creating a sustainable and productive homestead that brings joy and abundance for years.

Essential Tools and Equipment

You'll need more than enthusiasm to transform your backyard into a thriving homestead. The right tools and equipment are essential to your success, acting as extensions of your hands and embodying your intentions. This section will guide you through the essential tools and equipment every backyard homesteader should have, ensuring you're well-equipped to turn your vision into reality.

Quality gardening tools are non-negotiable. A sturdy, sharp spade or shovel is indispensable for breaking ground, turning soil, and planting. Look for one with a comfortable handle and a durable blade. A good set of hand tools, including a trowel, pruning shears, and a garden fork, will also be invaluable for day-to-day tasks such as planting, weeding, and harvesting.

Watering equipment is another critical component. Depending on the size of your homestead, this might range from a simple watering can for small plots to a more sophisticated drip irrigation system for larger areas. The goal is to ensure your plants receive the hydration they need with minimal waste and effort.

For those planning to incorporate livestock into their homestead, the proper housing and fencing are crucial. Chickens require a secure coop to protect them from predators, while larger animals like goats or pigs will need sturdy fencing to contain them. Additionally, consider your animals' feeding and watering systems to ensure they have constant access to fresh water and food.

Composting is a key aspect of sustainable homesteading. It turns kitchen scraps and yard waste into rich soil. Investing in a compost bin or setting up a compost pile will help you manage organic waste and improve your soil's health over time.

Lastly, protective gear should not be overlooked. Gardening gloves, sturdy boots, and a wide-brimmed hat can make your work more comfortable and prevent injuries. Remember, your safety and well-being are just as important as the tools you use.

As you gather these essential tools and equipment, remember that quality matters more than quantity. Choose items that are durable, comfortable to use, and suited to your specific needs. With the right tools, you'll be well on your way to creating a productive and sustainable backyard homestead.

Understanding Your Land and Climate

Backyard homesteading requires a deep understanding of your land and climate, as these factors are pivotal in shaping your homestead's success. This section delves into the essentials of getting to know your backyard's unique characteristics and how they influence the planning and execution of your homesteading projects.

Assessing the soil quality of your land is crucial. Soil types can vary dramatically, even within the same property, and understanding your soil's texture, pH, and fertility will guide you in selecting suitable crops and determining if amendments are necessary. Simple soil tests can be conducted using kits available at local gardening centers or through cooperative extension services, providing valuable insights into your soil's condition.

Next, consider the topography of your land. Your backyard's layout and elevation changes can influence water drainage, sunlight exposure, and wind patterns. Areas with southern exposure typically receive more sunlight, making them ideal for most vegetable and fruit crops. Conversely, low-lying areas might be prone to flooding or frost pockets and better suited for plants that thrive in wet conditions or for installing water management systems like rain gardens.

Climate plays a significant role in determining what you can grow and raise on your homestead. Understanding your region's hardiness zone will help you select plant varieties that thrive in your local conditions. Additionally, being aware of your area's average rainfall, temperature ranges, and seasonal changes will aid in planning your planting and harvesting schedules and preparing for extreme weather conditions.

Water availability is another critical aspect to consider. Whether relying on municipal water, a well, or rainwater collection systems, ensuring a reliable and sustainable water source is essential for irrigation, livestock, and household use. Assessing your water resources and planning for efficient water use can significantly impact your homestead's sustainability and productivity.

Lastly, observing your land's biodiversity can offer insights into creating a balanced and healthy ecosystem. Note the native plants and wildlife in your area, as they can indicate the health of your land and help you design your homestead to support local biodiversity. Encouraging beneficial insects, birds, and other wildlife can aid in pest control and pollination, enhancing your homestead's resilience and productivity.

By thoroughly understanding your land and climate, you can make informed decisions that align with the natural characteristics of your backyard, leading to a more productive and sustainable homestead. This foundational knowledge sets the stage for setting realistic goals and planning your homesteading projects confidently and clearly.

Setting Realistic Goals

After understanding your land and climate, the next step in establishing a thriving backyard homestead is to set realistic goals. This process is crucial, as it lays the foundation for what you aim to achieve with your space, resources, and time. It's about balancing dreams with practicality, ensuring that your homesteading efforts are both rewarding and sustainable.

First, assess your motivations for starting a backyard homestead. Are you looking to produce most of your own food, or are you more interested in the educational aspect of gardening and animal care for yourself or your family? Perhaps your goal is to reduce your carbon footprint or to create a more resilient lifestyle. Understanding your primary motivations will help guide your decisions moving forward.

Next, consider the resources you have available. This includes your physical space and climate, which you've already begun to understand, and your time, skills, and financial resources. Be honest about how much time you can dedicate to homesteading activities daily or weekly. Tasks like watering, weeding, feeding animals, and harvesting can become time-consuming, especially during peak seasons.

Financial resources are another critical factor. Start-up costs for a backyard homestead can vary widely depending on your goals. Simple gardening might require only seeds, soil, and essential tools, while more ambitious projects like keeping livestock or installing greenhouses can become quite costly. Plan a budget that reflects your goals and leaves room for unexpected expenses.

Skill level is another important consideration. If you're new to gardening, animal husbandry, or other homesteading activities, you may need to invest time in learning before you dive in. Fortunately, abundant resources are available, from books and online courses to local workshops and community gardens. Set goals that allow for a learning curve, starting with more straightforward projects and gradually working on more complex ones.

Finally, set specific, measurable, achievable, relevant, and time-bound (SMART) goals. Instead of a vague goal like "grow a garden," aim for something more concrete, such as "grow three types of vegetables to harvest by the end of the season." This approach will help you plan more effectively and track your progress, making necessary adjustments.

Remember, backyard homesteading is a journey, not a race. It's about learning and growing along with your garden and livestock. Setting realistic goals lays the groundwork for a fulfilling and sustainable homesteading experience that aligns with your lifestyle, resources, and aspirations. As you move forward, keep these goals in mind, but also be flexible and open to adjusting them based on what you learn and experience. The world of backyard homesteading is rich with opportunities for growth, discovery, and connection to the land.

1

SOIL AND COMPOSTING

A vibrant garden bathed in sunlight.

The Basics of Soil Health

Understanding the basics of soil health is crucial for any backyard homesteader. Soil isn't just dirt; it's a living, breathing entity that sustains your garden. Healthy soil is teeming with microorganisms, insects, and organic matter that work together to nourish the plants you grow. This section will guide you through the foundational knowledge needed to cultivate and maintain fertile soil in your homestead.

Soil health hinges on its structure and composition. Soil comprises minerals, organic matter, water, and air. The balance of these components determines the soil's texture and ability to retain water and nutrients. For instance, clay soils are nutrient-rich but often suffer from poor drainage, while sandy soils drain well but can struggle to hold onto nutrients and moisture.

To assess your soil's health, start with a simple texture test. Take a handful of moist soil and try to form it into a ball, then a ribbon. This tactile test can help you gauge whether your soil leans more towards clay, sand, or loam, which is the ideal balance of sand, silt, and clay. Loamy soil is the gold standard for gardeners, offering a perfect drainage and nutrient retention mix.

The pH level of your soil also plays a pivotal role in plant health. Most vegetables and fruits thrive in slightly acidic to neutral soil (pH 6.0-7.0). You can easily test your soil's pH with a kit from your local garden center. If your soil is too acidic or alkaline, it can be amended with lime or sulfur to bring it into the optimal range for your crops.

Organic matter is the lifeblood of healthy soil. It improves soil structure, aids in moisture retention, and provides a slow-release source of nutrients as it decomposes. Incorporating compost, aged manure, or leaf mold into your soil boosts its organic matter content and fosters a vibrant ecosystem below the surface.

Finally, understanding the importance of a living soil ecosystem is critical. A diverse microbial population aids in breaking down organic matter, making nutrients available to plants, and protecting against pests and diseases. Practices such as minimal tillage, cover cropping, and crop rotation support this underground community, enhancing soil health and fertility over time.

By grasping these basics of soil health, you lay the groundwork for a thriving backyard homestead. Healthy soil leads to robust plants, reduced pest and disease issues, and bountiful harvests. With this foundation, we can now focus on creating and maintaining a compost system, an essential practice for enriching your soil and closing the loop on your homestead's organic waste.

Creating and Maintaining a Compost System

Understanding the intricacies of creating and maintaining a compost system is a pivotal step in the journey toward a sustainable backyard homestead. Composting, the process of recycling organic matter into a rich soil amendment, is not just about waste reduction; it's about nurturing the soil that feeds us. This section delves into the practicalities of establishing a thriving compost system, offering a guide to transforming kitchen scraps and yard waste into black gold for your garden.

To begin with, select a suitable location for your compost pile or bin. It should be easily accessible but not too close to your living spaces to avoid potential odors. Consider the convenience of adding materials and turning the compost. A partially shaded spot is ideal, as it prevents the compost from drying out too quickly in the sun and keeps it warm enough to decompose in cooler weather.

There are various composting methods, depending on your

space and needs. A simple heap in a garden corner works for some, while others may prefer a compost bin to keep things tidy. Tumbler bins are excellent for those with limited space, as they facilitate easy turning and aerate the compost well. Whichever method you choose, ensure good air circulation and drainage to prevent the compost from becoming too wet or compacted.

The key to successful composting lies in the balance of 'greens' and 'browns'—the nitrogen-rich and carbon-rich materials, respectively. Greens include kitchen scraps like vegetable peels, coffee grounds, and fresh plant material, while browns comprise dried leaves, straw, and shredded paper. A general rule of thumb is maintaining a ratio of about 2:1, browns to greens, to ensure a healthy decomposition process without attracting pests or creating unpleasant odors.

Regular maintenance of your compost pile is crucial. Turning the compost every few weeks introduces oxygen, essential for the aerobic bacteria responsible for breaking down the materials. This also helps to distribute moisture and heat evenly throughout the pile, speeding up the decomposition process. If the compost seems too dry, adding water can help, but be cautious not to overwater.

Monitoring the temperature of your compost pile can provide valuable insights into its health. A well-functioning compost will heat up as the materials break down, often reaching temperatures between 130°F to 160°F. This heat is beneficial as it kills weed seeds and harmful pathogens. If the pile isn't heating up, it may need more greens, water, or aeration.

Finally, knowing when your compost is ready to use is essential. Mature compost is dark, crumbly, and has an earthy smell. It should no longer resemble the original materials but look like rich soil. This process can take several months to a year, depending on the conditions and materials used.

Incorporating compost into your garden beds enhances soil structure, moisture retention, and nutrient content, promoting healthy plant growth. It's a testament to the cycle of life in your backyard homestead, turning what was once considered waste into a valuable resource. As we move forward, understanding how to further enrich this compost with natural fertilizers and amendments will ensure your garden thrives in harmony with nature.

Natural Fertilizers and Amendments

Building on the foundation of a well-maintained compost system, it's essential to delve into the world of natural fertilizers and amendments to enrich your backyard homestead's soil further. These natural inputs are pivotal in creating a thriving, sustainable garden ecosystem. Unlike synthetic fertilizers, natural options feed your plants, improve soil structure, enhance microbial life, and reduce environmental impact.

One of the most straightforward and beneficial natural fertilizers is compost tea. Made by steeping finished compost in water, this nutrient-rich liquid can be applied directly to the soil or used as a foliar spray. It's an excellent way to give your plants a quick nutrient boost while introducing beneficial microorganisms to the soil.

Another valuable amendment is worm castings. Worms are nature's tillers and nutrient providers. Their castings are incredibly rich in essential plant nutrients, and incorporating them into your soil can significantly enhance plant growth and soil health. You can produce your worm castings by using kitchen scraps and garden waste by setting up a simple vermicomposting system.

Green manures, or cover crops, are also integral to natural soil fertility. Although they will be discussed in more detail in the following section, it's worth noting here that incorporating green manures into your garden rotation can significantly improve soil structure, fertility, and organic matter content. They are particularly beneficial for fixing nitrogen in the soil, a crucial nutrient for plant growth.

There are several natural options for those looking to add specific nutrients to their soil. Bone meal is an excellent source of phosphorus and vital for root development, while blood meal is a high-nitrogen amendment perfect for leafy growth. Wood ash can be lightly applied to the soil for potassium, which supports overall plant health and disease resistance. However, using these amendments judiciously is essential, as over-application can disrupt soil balance.

Seaweed, fresh, dried, or in a liquid extract form, is another fantastic soil amendment. It's a rich source of micronutrients and contains growth hormones and stimulants that can enhance plant health and productivity. Seaweed also increases plant stress tolerance, making your garden more resilient to drought, disease, and pests.

Lastly, rock dust is an often-overlooked amendment that can add a broad spectrum of minerals to the soil. Derived from finely ground rocks, it slowly releases essential minerals that plants need for growth. This is particularly beneficial for replenishing soils depleted over years of gardening.

Incorporating these natural fertilizers and amendments into your soil management practices can significantly enhance the health and productivity of your backyard homestead. By doing so, you provide your plants with the nutrients they need and contribute to a more sustainable and environmentally friendly gardening approach. As we move forward, understanding the role of cover crops and crop rotation will complement these practices, creating a holistic approach to soil health and fertility.

Cover Crops and Crop Rotation

Understanding the symbiotic relationship between cover crops and crop rotation is paramount to transforming your backyard into a thriving homestead. This section delves into how these practices nourish and protect your soil and set the stage for a bountiful harvest.

Cover crops, often called "green manure," are planted not for consumption but to cover the soil. They play a crucial role in enhancing soil health through various means:

- They prevent soil erosion by shielding the soil from the direct impact of raindrops.
- They suppress weeds by outcompeting them for sunlight and nutrients, reducing the need for chemical herbicides.
- Cover crops enhance soil fertility by fixing atmospheric nitrogen, mainly when using leguminous plants like clover or vetch.

When these crops are cut down and left on the surface, they decompose, adding organic matter and nutrients to the soil, thus improving its structure and water retention capacity.

Crop rotation, on the other hand, is the practice of growing different types of crops in the same area across a sequence of growing seasons. It reduces the reliance on chemical fertilizers and pesticides by naturally breaking cycles of pests and diseases. Each crop type absorbs specific nutrients from the soil, and by rotating them, you help prevent nutrient depletion. For instance, following a nitrogen-fixing legume crop with a nitrogen-loving leafy vegetable can optimize the nutrient use efficiency. Crop rotation also diversifies the soil microbiome, which is essential for a healthy ecosystem.

Integrating cover crops into your crop rotation plan can magnify these benefits. After harvesting your main crop, planting a cover crop before the next crop in the rotation takes over can keep the soil covered and active. This prevents nutrient leaching during off-seasons and ensures that the soil is revitalized and ready for the next planting cycle. For example, you might plant a winter cover crop like rye after harvesting tomatoes to protect and enrich the soil before planting spring vegetables.

To implement these practices effectively, start by assessing your soil's current condition and the specific needs of your homestead. Consider climate, soil type, and the crops you wish to grow. Plan your crop rotation schedule by grouping plants with similar nutrient needs or pest issues, and decide on the appropriate cover crops to plant between these groups. Remember, the goal is to maintain a continuous cycle of growth, rest, and rejuvenation in your soil, mimicking natural ecosystems.

In conclusion, integrating cover crops and crop rotation into your backyard homestead is a testament to the power of working with nature rather than against it. These practices are not only beneficial for your soil and crops but also contribute to a larger ecological balance. By adopting these methods, you're taking significant steps towards sustainability and resilience in your homesteading journey.

Testing and Understanding Your Soil

Understanding the soil in your backyard is akin to getting to know a close friend. It requires patience, observation, and a bit of science to comprehend its needs, strengths, and weaknesses truly. This understanding is crucial for any homesteader cultivating a thriving garden or farm. Soil testing is the first step in this journey of acquaintance, providing a baseline from which to work and improve your soil's health.

Soil testing is more straightforward than it might seem. It involves collecting soil samples from various parts of your homestead and sending them to a local extension service or a private lab for analysis. These tests offer invaluable insights into your soil's pH level, nutrient content (such as nitrogen, phosphorus, and potassium), and the presence of organic matter. Understanding these components helps you make informed decisions about soil amendments and select the right plants for your garden.

The pH level is critical to soil health and influences plant nutrient availability. Most vegetables thrive in slightly acidic to neutral soil (pH 6.0-7.0). If your soil is too acidic or alkaline, it can be amended with lime or sulfur to bring it to an optimal pH range. However, it's essential to follow the recommendations from your soil test report to avoid over-amendment.

Nutrient levels are equally important. Nitrogen, phosphorus, and potassium are plants' primary nutrients for growing. If your soil test indicates a deficiency in any of these, you can address it by adding organic or synthetic fertilizers. Yet, the beauty of a backyard homestead lies in the potential to use compost, manure, and other organic matter to enrich the soil, thus reducing the need for synthetic inputs.

Organic matter content is another crucial piece of the puzzle. It improves soil structure, water retention, and nutrient availability. Composting is a homesteader's best friend in this regard, transforming kitchen scraps and yard waste into black gold that nourishes the soil and supports a vibrant ecosystem underground.

Once you've tested your soil and understood its characteristics, the next step is continuously improving its health. This doesn't happen overnight but is a gradual process involving regular amendments, crop rotation, and adding organic matter. Remember, healthy soil is the foundation of a productive homestead. By nurturing the soil, you're not just growing plants but cultivating a sustainable future.

In summary, testing and understanding your soil is fundamental to backyard homesteading. It informs your decisions, guides your gardening practices, and ultimately leads to a more fruitful and fulfilling homesteading experience. With patience and persistence, you'll see your efforts reflected in the health of your soil and the abundance of your harvests.

Mulching Techniques

Having explored the intricacies of testing and understanding your soil, it's time to delve into mulching techniques—a critical next step in nurturing and protecting your garden. Mulching is not just about making your garden look neat; it is crucial in maintaining soil health, conserving moisture, and suppressing weeds. Let's explore how you can effectively implement mulching in your backyard homestead.

Mulching involves covering the soil surface around your plants with a material layer—organic or inorganic. This practice offers numerous benefits, including temperature regulation, moisture retention, and the reduction of soil erosion. Moreover, organic mulches contribute to soil fertility as they decompose, enriching the soil with essential nutrients.

Organic mulches are derived from natural materials that decompose over time. Some popular options include straw, grass clippings, leaves, wood chips, and compost. Each type has its unique benefits and considerations:

- **Straw and Grass Clippings:** These are readily available and excellent for vegetable gardens. They decompose quickly, adding organic matter to the soil. However, ensure that the grass isn't treated with herbicides, which could harm your plants.
- **Leaves:** An abundant resource in the fall, shredded leaves can be spread around your plants to provide insulation and gradually enrich the soil as they break down.
- **Wood Chips:** Ideal for perennial beds and pathways, wood chips last longer than most organic mulches but decompose slowly, adding less organic matter to the soil in the short term.
- **Compost:** Using compost as mulch not only suppresses weeds but also adds significant nutrients to the soil. It's particularly beneficial for improving soil structure and fertility.

Inorganic mulches include materials like plastic, landscape fabric, and gravel. While they don't improve soil fertility, they are effective in weed suppression and moisture conservation.

- **Plastic Mulch:** Commonly used in vegetable gardens, plastic mulch warms the soil and is excellent for heat-loving crops. However, it prevents water and air from reaching the soil and can be challenging to manage.
- **Landscape Fabric:** This porous material allows water and air to reach the soil while suppressing weeds. It's often used under gravel or other inorganic mulches to prevent them from sinking into the soil.
- **Gravel and Stones:** These materials are suitable for pathways and decorative purposes. They offer excellent weed control but can make the soil beneath them hotter.

When applying mulch, a few practical tips can ensure its effectiveness:

- **Depth:** Organic mulches should be layered 2-4 inches deep. Too much can suffocate plants, while too little may not effectively suppress weeds or retain moisture.
- **Timing:** Early spring is a great time to apply mulch after the soil has warmed up. Mulching too early can delay soil warming and plant growth.
- **Maintenance:** Refresh organic mulches to maintain the desired depth as they decompose. Keep inorganic

mulches clean and free from debris to prevent soil compaction.

Incorporating mulching into your gardening practices is a straightforward yet impactful way to enhance the health and productivity of your backyard homestead. Choosing the right mulch and applying it thoughtfully can create a more resilient and vibrant garden ecosystem.

Chapter Summary

- Soil health is fundamental for backyard homesteading, requiring a balance of minerals, organic matter, water, and air.
- Soil texture and pH level are critical; loamy soil and a slightly acidic to neutral pH (6.0-7.0) are ideal for most plants.
- Organic matter, like compost and aged manure, is vital for improving soil structure and fertility.
- A diverse microbial population in the soil supports plant health by breaking down organic matter and protecting against pests.
- Composting kitchen scraps and yard waste enriches soil and reduces waste, and a balance of greens and browns is necessary for successful decomposition.
- Natural fertilizers and amendments, such as compost tea, worm castings, and green manures, enhance soil without the environmental impact of synthetic options.

- Cover crops and crop rotation improve soil health by preventing erosion, suppressing weeds, fixing atmospheric nitrogen, and breaking pest cycles.
- Mulching with organic or inorganic materials conserves moisture, suppresses weeds, and can improve soil fertility as organic mulches decompose.

2

GARDENING

A person in a wide-brimmed hat tending to a garden.

Planning Your Garden

Planning your garden is akin to painting a canvas, where the soil is your medium, and the plants are your palette. The key to a successful garden, especially in the context of a backyard homestead, lies in meticulous planning and a deep understanding of the space you have at your disposal. This section will guide you through the essential steps to plan your garden efficiently, ensuring a bountiful harvest and a vibrant oasis in your backyard.

First and foremost, assess the space available to you. Not all homesteads are blessed with expansive yards, but even the smallest spaces can be transformed into productive gardens with the right approach. Observe the patterns of sunlight and shade throughout the day, as this will significantly influence what you can grow and where. To thrive, most vegetables and fruits require at least six hours of sunlight daily, so plot your garden accordingly.

Next, consider the soil quality in your chosen garden area. Soil health is paramount in organic gardening, and a backyard homestead relies heavily on the sustainability of its resources. Conduct a soil test to understand its composition, pH level, and nutrient profile. This information will guide you in amending your soil, ensuring it provides the perfect plant foundation. Composting is a fantastic way to enrich your soil naturally, turning kitchen scraps and yard waste into gold for your garden.

Water access is another critical factor in garden planning. Efficient water use conserves this precious resource and promotes healthier plants. Explore options like rainwater harvesting or setting up a drip irrigation system to provide a consistent and measured amount of water to your plants directly at their roots, where it's most needed.

Now that you understand your space, soil, and water setup, it's time to think about your garden's layout. Raised beds, container gardening, and traditional in-ground plots each have advantages and can be mixed to suit your needs and preferences. Raised beds, for example, offer excellent drainage and can help deter some pests, while container gardening allows for greater flexibility in managing sunlight exposure.

As you sketch out your garden plan, consider the importance of crop rotation and companion planting. These practices not only maximize space and yield but also contribute to the health of your garden by preventing soil depletion and reducing pest and disease issues. Integrate flowers and herbs to attract pollinators and beneficial insects, further enhancing the ecosystem of your backyard homestead.

In planning your garden, remember that flexibility and observation are your allies. Nature often reminds us of its unpredictability, so be prepared to adapt your plans as needed. Keep a garden journal to record what works and what doesn't, setting the stage for continuous learning and improvement in your gardening journey.

By following these steps, you're not just planning a garden; you're laying the groundwork for a sustainable, productive, and beautiful extension of your home. The effort you put into planning today will pay dividends in the coming seasons, providing nourishment and joy for your household.

Choosing the Right Plants

After carefully planning your garden, considering the layout, sunlight, and soil conditions, the crucial next step in your backyard homesteading journey involves selecting the right plants for your space. This decision is vital for a thriving garden and requires a thoughtful approach to match your garden's conditions with the plants' needs.

First, it's essential to understand your climate zone, as this knowledge will guide you in choosing plants well-suited to your area's weather patterns and temperature ranges. Each plant has specific climate preferences; selecting those adapted to your zone increases the likelihood of success.

Reflect on the sunlight and soil assessment you conducted during the planning phase. Different plants have varying requirements for sunlight, ranging from full sun to partial shade, and soil type and pH level can significantly affect plant growth. Choose plants that will thrive in the conditions you have to offer. Decide on the types of plants you want to grow, whether aiming for a vegetable garden, an herb garden, or a mix of both. To enhance your homestead's aesthetics, you may also be interested in fruit trees or ornamental flowers, each with considerations such as the space and care needed.

Think about the growing seasons of the plants you're considering. Some plants are perennials, returning year after year, while others are annuals or biennials, requiring replanting. Planning for a mix can ensure your garden remains productive and vibrant across different seasons. Evaluate the space requirements and growth habits of potential plants. Some plants, like squash, need much room to sprawl, while others, such as tomatoes, grow upwards and can be supported with stakes or cages. Understanding these habits will help you maximize your garden space and avoid overcrowding.

Consider the benefits of companion planting, where some plants, when grown together, can improve each other's health and yields. For example, marigolds repel certain pests and can be a great companion for many vegetable plants. Researching companion planting combinations can lead to a more harmonious and productive garden. Finally, think about your personal preferences and goals. Choose plants you and your family enjoy eating or that fulfill your desired purpose, whether cooking, herbal remedies, or creating a beautiful space.

Your garden should reflect your interests and needs, making the experience enjoyable and rewarding. By carefully selecting the right plants for your garden, you're setting the stage for a successful and fulfilling gardening season, looking forward to a bountiful harvest and the many joys of backyard homesteading.

Seed Starting and Transplanting

Cultivating a thriving backyard homestead garden begins with the pivotal steps of seed-starting and transplanting. These steps not only allow for a deeper connection with your plants but also provide the opportunity to kickstart your garden with a variety of species that may not be readily available as seedlings in your local area.

The adventure starts indoors, where seeds are given a controlled environment to sprout and grow into healthy seedlings. It's essential to select high-quality seeds and suitable containers with adequate drainage, such as recycled containers, peat pots, or commercially available seed trays, and fill them with a sterile, seed-starting mix to prevent disease. Maintaining moisture and warmth, around 65-75°F (18-24°C), is crucial for encouraging germination, and a plastic cover or dome can help retain humidity. However, it should be removed once seeds sprout to prevent mold growth.

Once seeds have germinated, ensuring they receive at least 16 hours of light daily is vital to prevent them from becoming leggy and weak. As seedlings grow, it's important to harden them off by gradually exposing them to outdoor conditions over a week to reduce transplant shock and acclimate them to their new environment.

When it's time to transplant, choosing a cool, overcast day can ease the transition for young plants. The garden soil should be well-prepared, enriched with compost, and moist. Seedlings should be carefully removed from their containers, handled by the leaves to avoid stem damage, and placed in a hole big enough to accommodate the root ball. The soil should be gently firmed around them to eliminate air pockets. Watering thoroughly after planting is crucial for root establishment.

Spacing is critical to prevent overcrowding and ensure each plant has enough room to grow and access nutrients. Specific spacing recommendations are usually found on the seed packet or plant tag. After transplanting, it's important to keep a close eye on the seedlings, especially during the first few weeks, to ensure they adapt well to their new environment and provide support structures for climbing plants early on to avoid disturbing the roots later.

In conclusion, seed starting and transplanting are rewarding steps in the gardening process that pave the way for a bountiful harvest, and by nurturing your plants with patience and care, you'll cultivate a vibrant garden and a deeper appreciation for the cycle of growth and renewal in your backyard homestead.

Watering and Irrigation

Watering and irrigation are the lifelines of a thriving garden. After successfully starting your seeds and transplanting them into the garden, the next critical step is ensuring they receive the right amount of water. This section will guide you through the essentials of watering and irrigation, providing your plants with the best chance for growth, health, and productivity.

Understanding your plants' needs is the first step in effective watering. Different plants have varying water requirements, and even the time of day you water can significantly impact their health. Early morning is generally the best time to water your garden. It allows the water to reach deep into the soil, encouraging deep root growth while minimizing evaporation and the risk of fungal diseases that can occur with evening watering.

There are several methods of watering, each with its own set of benefits. Hand watering with a hose or watering can is the most straightforward method, allowing for direct control over the amount of water each plant receives. However, it can be time-consuming and is not always the most efficient method for larger gardens.

Drip irrigation systems are a more efficient alternative, delivering water directly to the base of each plant. This method reduces water waste and helps keep the leaves dry, crucial in preventing many common plant diseases. Setting up a drip irrigation system might require an initial investment of time and resources, but the long-term benefits of water conservation and healthier plants are well worth it.

For those looking for a more automated solution, soaker hoses and sprinkler systems can cover larger areas with minimal effort. Like drip irrigation, soaker hoses allow water to seep slowly along their length, providing a steady supply of moisture to the plant roots. Sprinkler systems, on the other hand, can water a large area from above, simulating rainfall. While sprinklers are less targeted than drip or soaker systems, they can effectively water lawns or large garden areas.

No matter your chosen method, monitoring your garden's moisture levels is crucial. Over-watering can be as harmful as under-watering, leading to root rot and other issues. A simple way to check soil moisture is to insert a finger into the soil near your plants; if the soil feels dry at a depth of about an inch, it's time to water.

Incorporating mulch around your plants can also aid in moisture retention, reducing the frequency of watering needed. Mulch acts as a barrier, slowing evaporation and contributing to soil health as it breaks down over time.

As we progress in our gardening journey, understanding the balance and techniques of watering and irrigation sets the stage for a healthy, productive garden. This knowledge conserves water and ensures that our plants have the resources they need to thrive. With our watering strategies in place, we can turn our attention to the next critical aspect of gardening: managing pests and diseases to protect our hard-earned harvests.

Pest and Disease Management

A flourishing garden is a source of pride and sustenance in backyard homesteading. However, this green paradise can quickly become a battleground where pests and diseases threaten to overrun your hard work. Effective management of these unwelcome visitors is crucial to ensure the health and productivity of your garden. This section delves into practical strategies for keeping pests and diseases at bay, ensuring your garden remains a thriving part of your homestead.

Understanding the enemy is the first step in effective pest and disease management. Common garden pests include aphids, caterpillars, slugs, and beetles, each with their preferred plants and attack methods. On the other hand, diseases can be fungal, bacterial, or viral, with symptoms ranging from mildew to leaf spots and wilting. Regularly monitoring your garden will help you identify problems early, which is key to controlling outbreaks.

Prevention is always better than cure. Maintaining healthy soil through composting and crop rotation supports strong plant growth and is less susceptible to pests and diseases. Choosing disease-resistant plant varieties can also significantly reduce the risk of outbreaks. Additionally, encouraging natural predators into your garden, such as ladybugs, birds, and frogs, helps keep pest populations in check.

When intervention is necessary, opting for organic and natural remedies is advisable to maintain the balance of your backyard ecosystem. Neem oil, diatomaceous earth, and insecticidal soaps are effective against many pests without harming beneficial insects. Similarly, baking soda, copper fungicides, and sulfur can organically treat many common plant diseases.

Sometimes, despite all efforts, pests or diseases may establish a foothold in your garden. In such cases, removing and destroying infected plants or parts of plants can prevent the spread of disease. For severe pest infestations, manual removal or the use of pheromone traps may be necessary.

Remember, pest and disease management aims not to create a sterile environment but to maintain a healthy balance that allows your garden to thrive. By adopting a proactive and mindful approach, you can protect your garden from pests and diseases, ensuring it remains a vibrant and productive part of your backyard homestead.

Harvesting and Storing Your Produce

Diligently managing pests and diseases in your garden leads to the rewarding phase of harvesting and storing your produce, which is crucial for maximizing the fruits of your labor, ensuring that nothing goes to waste, and enjoying your garden's bounty long after the growing season.

The key to successful harvesting is timing, as most vegetables and fruits have a prime period when their flavor and nutritional content peak. For instance, leafy greens are best harvested in the morning when their moisture content is highest, while root vegetables like carrots and beets should be picked when they reach a desirable size.

Tomatoes, on the other hand, are ripe for picking when brightly colored and slightly firm. Using the right tools and techniques is essential to avoid damaging the plant or produce, with a sharp knife or gardening shears for clean cuts and gentle handling to prevent bruising.

Once harvested, proper storage is essential to preserve the freshness and flavor of your produce. Different fruits and vegetables have varying storage needs.

Root vegetables like potatoes and onions prefer cool, dark, and dry places and can last for months under the right conditions, typically in a cellar or cool pantry. Leafy greens, however, need refrigeration and a moisture-retaining bag to stay fresh.

Fruits like apples and pears can be stored in a cool, humid environment for an extended period. Still, it's essential to be mindful of ethylene-producing fruits like apples, which can hasten the ripening and spoiling of other produce if stored too closely together.

For those with a surplus of produce, preserving methods such as canning, freezing, or drying can extend the life of your harvest and provide you with homegrown flavors even in the off-season.

Canning tomatoes, making jams from berries, freezing vegetables like beans, peas, and corn, and drying herbs and some fruits are excellent ways to preserve the nutritional value and taste of your garden's bounty.

Harvesting and storing produce is a way to extend its shelf life while maintaining its nutritional value and taste. This allows you to enjoy the fruits of your labor throughout the year and makes your backyard homestead a truly rewarding endeavor.

Chapter Summary

- Planning a garden involves assessing space, understanding soil quality, and ensuring water access for a successful harvest.
- The garden layout, including raised beds and container gardening, should consider sunlight exposure and incorporate crop rotation and companion planting.
- Choosing the right plants for your garden requires understanding your climate zone, assessing sunlight and soil, and considering plant types and their seasonal growth.
- Seed starting indoors involves selecting quality seeds, providing moisture and warmth, and ensuring adequate light. Then, with proper spacing and care, the seeds are transplanted into the garden.

- Effective watering and irrigation methods, like drip systems or soaker hoses, are crucial for plant health, with early morning watering being optimal.
- Pest and disease management includes regular monitoring, using organic remedies, and encouraging natural predators to maintain a healthy garden ecosystem.
- Harvesting produce at the right time maximizes flavor and nutritional content, with proper storage methods extending freshness and usability.
- Preserving surplus produce through canning, freezing, or drying ensures a year-round supply of homegrown fruits, vegetables, and herbs.

3

RAISING CHICKENS

A chicken coop standing in a vast field.

Choosing the Right Breeds

Raising chickens in your backyard homestead brings the exciting task of choosing the right breeds to suit your needs and environment. This decision is crucial, as the breeds you select will directly impact your homesteading experience, from the quantity and quality of eggs produced to the temperament and adaptability of the chickens to your specific climate.

Firstly, consider what your primary goal is in raising chickens. Are you looking for prolific egg layers or are you more interested in meat production? You may be seeking dual-purpose breeds that are capable of providing both. For those focused on egg production, breeds like the Leghorn or Rhode Island Red are renowned for their high yield of eggs. On the other hand, if meat production is your goal, breeds such as the Cornish Cross grow quickly and provide a substantial amount of meat. For homesteaders looking for versatility, the Plymouth Rock and Sussex breeds balance egg production and meat quality.

Another important factor to consider is the climate of your homestead. Some chicken breeds are more resilient to cold weather, such as the Buff Orpington and the Barred Rock, known for their thick feathering that provides insulation during colder months. Conversely, if you live in a warmer climate, breeds like the Australorp or the Welsummer, which are more tolerant of heat, might be more suitable.

Temperament is another critical consideration. If you're raising chickens in a family setting, you might prefer breeds known for their docile and friendly nature, such as the Silkie or the Cochin. These breeds are less likely to be aggressive and can become affectionate pets.

Lastly, think about the space you have available. Some breeds, like the Bantam varieties, are smaller and require less space, making them ideal for smaller backyards. Meanwhile, larger breeds will need more room to roam and forage.

In choosing the suitable breeds for your backyard

homestead, it's essential to do thorough research and consider all these factors. By aligning your choices with your goals, climate, space, and lifestyle, you'll ensure a rewarding and sustainable chicken-raising experience. Remember, each breed has unique characteristics and needs, so take the time to understand what each can bring to your homestead. With careful selection, your chickens will thrive and become a cherished part of your homesteading journey.

Housing and Coop Design

After selecting the breeds that best suit your backyard homestead, it's essential to focus on their future home. A well-designed chicken coop is crucial for the shelter, health, and happiness of your chickens.

The space inside the coop should be at least 3-4 square feet per chicken, with about 10 square feet in the outdoor run to prevent overcrowding, stress, pecking, and the spread of diseases. For a flock of 6 chickens, a coop of at least 18-24 square feet with an attached run of 60 square feet is recommended.

Proper ventilation in the coop is necessary to remove moisture and ammonia fumes, keeping the air fresh and reducing the risk of respiratory issues. However, the coop also needs to be insulated enough to keep the chickens warm in colder climates while allowing air circulation. This can be achieved with strategically placed vents or windows.

The coop must also be secure against predators, using durable hardware cloth for enclosures and burying it at least 12 inches underground to deter digging predators. Lockable doors and windows will enhance security.

Nesting boxes should be provided for hens to lay their eggs, with one box for every three to four hens, placed in a quiet, darker part of the coop. Roosting bars, positioned higher than the nesting boxes, should accommodate all chickens comfortably, aligning with their instinct to roost in high places.

The coop design should also facilitate easy cleaning to maintain health, including removable trays and hose-down floors. Accessibility is important for egg collection, refilling feeders and waterers, and performing cleaning and maintenance tasks, so ensure the door is large enough for comfortable entry.

A well-thought-out coop design is vital for the well-being of your chickens, leading to a more productive and rewarding backyard homestead experience. With the coop set up, attention can then turn to feeding and nourishing your flock to keep them in peak condition.

Feeding and Nutrition

Transitioning from the essentials of housing and coop design, we delve into the equally critical aspect of raising chickens: **feeding** and **nutrition**. This section equips you with the knowledge to nourish your flock effectively, ensuring they are healthy, happy, and productive.

Chickens require a balanced diet consisting of carbohydrates, proteins, fats, vitamins, and minerals, usually formulated by commercial poultry feed to meet these needs. It's crucial to understand the components of their feed and any additional supplements they might need for their overall well-being.

Different feed types exist for different stages of a chicken's life. Starter feed, high in protein, is essential for chicks up to 6 weeks old to support their rapid growth. With slightly less protein, grower feed is for chickens from 6 weeks until they begin laying eggs.

Layer feed, rich in calcium, is necessary once hens start laying eggs to ensure strong eggshells and maintain their health. Scratch grains, a mix of grains, can be offered sparingly as a treat to encourage natural foraging behavior.

While commercial feeds are formulated to meet all nutritional requirements, offering supplements like oyster shells for additional calcium or grit to aid digestion can address specific needs or deficiencies. Treats can include fruits, vegetables, and grains but should be given in moderation to avoid nutritional imbalances and obesity. Ensuring chickens access fresh, clean water is vital for their digestion and overall health.

Feeding practices, such as feeding chickens at the same time each day, help establish a routine, reduce stress, and promote health. Monitoring feed intake to adjust portions as necessary and cleaning feeders regularly to prevent mold and contamination is essential for preventing health issues. Seasonal changes may affect chickens' nutritional needs, requiring more feed during colder months to stay warm and ensuring access to cool, clean water in summer to prevent dehydration.

Proper feeding and nutrition are foundational to raising a healthy, productive flock. By understanding and meeting your chickens' dietary needs, you ensure their well-being and the success of your backyard homestead. As we move on from feeding and nutrition, we'll explore health and wellness, ensuring your chickens remain in peak condition year-round.

Health and Wellness

Maintaining the health and wellness of your backyard chickens is not just a responsibility; it's necessary to ensure a thriving homestead. This section delves into practical strategies and insights to keep your feathered friends in peak condition, focusing on preventive care, recognizing signs of illness, and addressing common health issues.

Preventive care is the cornerstone of chicken health and wellness. Regularly scheduled check-ups are crucial, even for backyard flocks. These check-ups should include examining your chickens for signs of distress, injury, or illness. Vaccinations play a vital role in preventing diseases that can devastate unvaccinated flocks. Consult with a veterinarian with poultry experience to establish a vaccination schedule tailored to your specific needs and local disease risks.

Parasite control is another critical aspect of preventive care. External parasites like mites and lice can cause discomfort and health issues in chickens, while internal parasites like worms can lead to more severe health problems. Regular inspection and treatment for parasites are necessary to keep your chickens healthy. Various treatments are available, including natural remedies and chemical treatments, but choosing the method that best suits your homestead's philosophy and your chickens' needs is essential.

Recognizing the signs of illness early can be the difference between life and death for your chickens. Symptoms such as lethargy, reduced appetite, abnormal droppings, and changes in egg production can indicate health issues. Respiratory problems, often characterized by sneezing, coughing, or discharge from the nostrils, require immediate attention. Isolating sick birds from the rest of the flock is crucial to prevent the spread of disease.

Common health issues in backyard chickens include respiratory infections, digestive problems, and reproductive disorders. Many of these conditions are manageable with prompt and appropriate care. For instance, providing a clean, dry, and well-ventilated living environment can prevent many respiratory and digestive problems. Additionally, ensuring your chickens have a balanced diet of essential nutrients can help prevent nutritional deficiencies that lead to health issues.

Consulting with a poultry veterinarian is always the best course of action in cases of illness. They can provide accurate diagnoses and treatment options. Sometimes, despite our best efforts, chickens will fall ill. Knowing when to seek professional help is vital to responsible chicken care.

In conclusion, the health and wellness of your backyard chickens depend on a combination of preventive care, early detection of illnesses, and appropriate treatment. By staying informed and vigilant, you can ensure your chickens lead healthy, productive lives, paving the way for a successful transition into discussions about egg production and beyond. Remember, healthy chickens are happy chickens, and happy chickens are the heart of a thriving backyard homestead.

Egg Production

In backyard homesteading, egg production from your flock of chickens can be both a rewarding and practical aspect of self-sufficiency. After ensuring the health and wellness of your chickens, the next logical step is to optimize their environment and care for maximum egg production. This section will guide you toward a plentiful and consistent egg yield from your backyard flock.

First and foremost, the breed of chicken you choose plays a significant role in your egg production efforts. Some breeds, like the Leghorn, Australorp, and Rhode Island Red, are known for their prolific egg-laying abilities. When selecting your flock, consider the climate of your area and the breed's adaptability to ensure their comfort and productivity.

Nutrition is the cornerstone of healthy, productive chickens. A balanced diet is crucial for egg production. Layers require a diet rich in calcium, protein, and essential vitamins. Providing a quality layer feed is the most straightforward way to meet these nutritional needs. Additionally, supplementing their diet with kitchen scraps, garden waste, and a constant fresh water supply can boost their overall health and egg output.

Light exposure significantly influences a chicken's laying cycle. Chickens need about 14-16 hours of daylight to maintain optimal egg production. During shorter winter days, consider using a light in the coop to extend the amount of perceived daylight. However, balancing this with periods of natural darkness is essential to ensure the chickens' health and well-being.

The living conditions of your chickens also affect their ability to lay consistently. Ensure the coop is clean, well-ventilated, and protected from extreme weather conditions. Each chicken requires enough space to move freely, roost, and lay eggs without stress. Nesting boxes should be cozy, dark, and soft bedding to encourage laying. Typically, one nesting box for every three to four hens is sufficient.

Stress reduction is another critical factor in maximizing egg production. Chickens are sensitive to changes and can be stressed by loud noises, predators, and overcrowding. Maintaining a calm, safe environment and handling your chickens gently can help keep stress levels low and egg production high.

Regular health checks are vital to prevent diseases and parasites, which can significantly impact egg production. Implementing preventative health measures and addressing any issues promptly will keep your flock healthy and productive.

In conclusion, optimizing egg production in your backyard homestead involves:

- A combination of selecting the suitable breeds.
- Providing a balanced diet.
- Managing light exposure.
- Ensuring comfortable living conditions.
- Reducing stress.
- Maintaining good health practices.

By focusing on these areas, you can enjoy a steady supply of fresh, nutritious eggs from your backyard flock, contributing significantly to your homestead's self-sufficiency and sustainability.

Dealing with Predators

Discussing the intricacies of egg production naturally leads to a more challenging aspect of raising chickens: dealing with predators. This part of the guide is designed to equip you with the knowledge and strategies necessary to protect your flock from common threats and ensure their safety and well-being.

Predators can vary widely depending on location, but foxes,

raccoons, hawks, and neighborhood dogs are among the most common. Each predator has its method of attack, necessitating specific defensive strategies.

Understanding the behavior of potential predators is crucial to protecting your chickens effectively. For example, raccoons can open latches and climb over fences due to their dexterity, while hawks target free-ranging chickens from the sky. Tailoring your defenses to the specific threats posed by local predators is essential.

The first line of defense is a well-constructed coop and run. Ensuring your chicken coop is sturdy, with no gaps or weak points, is essential. Hardware cloth is recommended over the chicken wire for its durability and difficulty in tearing. Burying the wire at least a foot deep around the perimeter can deter digging predators like foxes.

Consider covering the run with a solid roof or securely attached netting for aerial threats, which protects chickens from hawks and owls without blocking sunlight.

Various deterrents can also help protect your flock. Motion-activated lights, sprinklers, and noise deterrents like radios left on at night can scare away nocturnal predators. Guard animals, such as dogs, geese, or llamas, can effectively deter predators, provided they coexist peacefully with your chickens.

Vigilance in protecting your chickens from predators involves regularly inspecting your coop and run for signs of attempted entry or damage and addressing any vulnerabilities immediately. Keeping the area around the coop clean and free of attractants, such as spilled feed, can reduce the likelihood of predators being drawn to your backyard.

By understanding local predator threats and implementing a combination of physical barriers, deterrents, and vigilant monitoring, you can create a safe environment for your chickens. Protecting your flock is an ongoing effort, but with the right strategies, you can minimize risks and enjoy the rewards of raising chickens in your backyard homestead.

Chapter Summary

- Choosing the right breed involves selecting chicken breeds for your backyard homestead by considering goals such as egg production, meat, or both, along with climate adaptability, temperament, and space requirements.
- Housing and coop design requires creating a healthy and safe environment for chickens with adequate space, ventilation, insulation, predator protection, nesting boxes, roosting bars, ease of cleaning, and accessibility.
- Feeding and nutrition involve providing a balanced diet with commercial poultry feed, supplements for specific needs, and clean water, adjusting feeding practices, and considering seasonal changes to maintain health and productivity.
- Health and wellness focuses on keeping chickens healthy through preventive care, recognizing signs of illness, managing parasites, consulting with a veterinarian, and promptly addressing common health issues.
- Egg production can be optimized by selecting the right breed, ensuring proper nutrition, light exposure, living conditions, stress reduction, and health

practices to maximize egg yield and contribute to homestead sustainability.

- Dealing with predators includes understanding predator behavior, fortifying the coop and run, implementing deterrents, and maintaining vigilance to protect chickens from common threats.
- The overall strategy for raising chickens emphasizes the importance of thorough planning and consideration in breed selection, coop design, nutrition, health management, egg production optimization, and predator protection for a successful backyard homestead.
- Sustainable homesteading highlights that raising chickens is a rewarding aspect of self-sufficiency that requires careful selection, proper care, and protection to ensure the well-being and productivity of the flock.

4

BEEKEEPING

A beekeeper in protective gear holding a box of honey.

Understanding Bee Behavior

Diving into the world of beekeeping, starting with a foundational understanding of bee behavior, is crucial. This knowledge enriches the beekeeping experience and ensures the safety and productivity of your beekeeping endeavors. Bees, fascinating creatures that they are, exhibit complex and highly organized behaviors, reflecting their roles within the hive and their interaction with the environment.

At the heart of bee behavior is the hive's social structure, which is dominated by the queen bee, whose primary role is reproduction. Worker bees, which are female, perform various tasks, including foraging for nectar and pollen, feeding the queen and larvae, and maintaining the hive's cleanliness and temperature. Drones, or male bees, solely aim to mate with a new queen. Understanding these roles is crucial for managing your hive effectively.

Communication among bees is another fascinating aspect. They use a combination of pheromones, or chemical signals, and the famous "waggle dance" to convey information about food sources, hive health, and other critical factors. Observing these communication methods can provide insights into the hive's needs and challenges.

Bees also exhibit defensive behaviors to protect the hive. They may sting when they perceive a threat to their hive or queen. To minimize stress on the bees and the beekeeper, it's essential to approach beekeeping with respect for these natural behaviors, using protective gear and learning how to interact with bees.

Understanding seasonal behaviors is equally important. Bees are active in warmer months, focusing on foraging and building up the hive's resources. In winter, they cluster around the queen, keeping her warm and focusing on survival. This seasonal cycle affects how you care for your bees, from feeding in the winter to managing the hive's growth in the summer.

Lastly, bees face various challenges, from diseases and pests to environmental stressors. Being attuned to your hive's signs of distress or disease can help you promptly address these issues.

A deep understanding of bee behavior is not just about becoming a successful beekeeper; it's about fostering a harmonious relationship with these incredible insects and contributing to their preservation. As we move forward, we'll delve into the practical steps of setting up your hive, building on the foundation of understanding we've established here.

Setting Up Your Hive

After gaining a foundational understanding of bee behavior, the next logical step is to set up your hive - a task that requires careful planning, respect for the bees, and a bit of elbow grease.

First and foremost, selecting the right location for your hive is paramount. Bees thrive in environments that provide ample sunlight, protection from strong winds, and easy access to water and foraging materials. An ideal spot is a south-facing spot that catches the morning sun yet is shaded during the hottest part of the day. Use a sturdy stand or concrete blocks to ensure the hive is elevated from the ground to protect it from moisture and predators.

Several types of hives exist when choosing your hive, including the traditional Langstroth hive, the top-bar hive, or the more recent Warre hive. Each has advantages and is designed to cater to different beekeeping philosophies and practices. The Langstroth, for instance, is widely used due to its practicality and ease of honey extraction. However, the choice ultimately depends on your preference, the amount of time you can dedicate to beekeeping, and the goals you wish to achieve with your hive.

Assembling your hive requires attention to detail. Whether starting with a pre-assembled kit or building from scratch, ensure all components fit snugly together to prevent drafts and protect your bees from the elements. The hive should consist of a bottom board, the hive boxes or supers where the bees will live and store honey, frames to build their comb, a queen excluder to keep the queen in the brood chamber, and a cover to seal the top.

Before introducing your bees to their new home, take the time to familiarize yourself with the tools of the trade. A bee suit, gloves, a smoker, and a hive tool are essential for handling your bees safely and effectively. The smoker calms the bees, making them easier to work with during hive inspections and maintenance.

Introducing bees to your hive is a thrilling moment. The process requires gentleness and patience, whether starting with a nucleus colony, a package of bees, or capturing a swarm. Open the hive and carefully place the frames inside, ensuring the queen is securely introduced. Once the bees are in, close the hive and give them time to acclimate to their new surroundings, resisting the urge to check on them too frequently in the first few days.

Setting up your hive is just the beginning of a fascinating journey into beekeeping. As your bees settle in, the focus shifts to maintaining the health and productivity of the hive, ensuring both your bees and your backyard homestead thrive in harmony. With dedication and care, your beekeeping endeavors will yield sweet rewards and contribute to the well-being of the local ecosystem, making your homestead a beacon of sustainability and biodiversity.

Maintenance and Hive Health

Once your hive is established, the ongoing journey of maintenance and ensuring hive health begins. This critical phase in beekeeping is about keeping your bees alive and helping them thrive. A healthy hive is a productive hive, and as a backyard beekeeper, your role shifts to that of a caretaker, closely monitoring and managing the well-being of your buzzing inhabitants.

Regular inspections are the cornerstone of hive maintenance. Aim to check your hive every two weeks during the active season. These inspections are vital for spotting early signs of disease, assessing the queen's health and productivity, and ensuring the hive has enough space to grow. When inspecting, look for a pattern of healthy brood (eggs, larvae, and pupae), which indicates a strong, laying queen. Also, watch for pests such as varroa mites or hive beetles, which can quickly bring down a healthy colony.

Varroa mites are among the most common and destructive pests beekeepers face. These tiny parasites attach themselves to bees and larvae, weakening and spreading diseases within the colony. Integrated pest management strategies, such as regular mite counts and natural miticides, can help keep these pests at bay. Remember, a proactive approach is always better than a reactive one regarding pest management.

Another aspect of hive health is ensuring bees access diverse and abundant food sources. Planting a bee-friendly garden or ensuring your bees can forage in a pesticide-free environment can significantly impact their health and productivity. Water is also essential, especially in hot weather, so providing a clean, accessible water source near the hive is crucial.

Swarming is a natural part of bee behavior, especially in strong, growing colonies. However, it can significantly reduce your hive's population and productivity. To manage swarming, consider splitting strong colonies or providing additional space and resources to accommodate your hive's growth.

Finally, preparing your bees for winter is an essential part of maintenance. This includes ensuring they have enough honey stores to last through the cold months, reducing the hive's entrance to protect against intruders, and providing insulation to help maintain a stable temperature within the hive.

By staying vigilant and responsive to your hive's needs, you can foster a thriving bee community in your backyard. Remember, beekeeping is a journey of learning and adaptation. Each hive has its unique challenges and rewards, and with patience and care, you'll be well on your way to a successful harvest.

Harvesting Honey

Harvesting honey is one of the most rewarding aspects of beekeeping, offering a sweet payoff for the care and effort you've invested in your hives. While enjoyable, this process requires careful timing and technique to ensure the health of your bee colony and the quality of the honey collected. Let's delve into the practical steps and considerations for a successful honey harvest.

Firstly, timing is crucial. Honey should be harvested at the end of the blooming period when bees have had ample time to gather nectar and convert it into honey. This typically occurs in late summer or early fall, depending on your local climate and flora. The honey must be fully mature, which bees indicate by capping the honey cells with wax. Harvesting too early can lead to honey that's too high in moisture and prone to fermentation.

Before you begin, ensure you have the right tools on hand. A bee suit, gloves, and a smoker are essential to protect yourself from stings and to calm the bees. You'll also need a hive tool to open the hive and separate the frames, a bee brush to remove any bees from the frames gently, and an extractor to spin the honey out of the comb.

The process starts with gently smoking the hive to calm the bees and reduce aggression. Carefully remove the lid and the inner cover, then gradually pull out the frames, inspecting them to ensure they're capped and ready for extraction. Working calmly and steadily is essential to minimize disturbance to the bees.

Once you've selected the frames to harvest, use the bee brush to sweep any remaining bees off the frames gently. It's crucial to do this gently to avoid harming the bees. Then, transport the frames to your extraction area, which should be set up in a clean, bee-proof space to avoid attracting bees or other insects.

The next step involves uncapping the wax seals on the honeycomb, either with a heated knife or a special uncapping fork. This exposes the honey, making it ready for extraction. Place the uncapped frames in the extractor, a centrifuge that forcefully spins the honey out of the comb. After extraction, the honey can be filtered to remove wax particles and stored in clean, airtight containers.

It's essential to leave enough honey in the hive to sustain the bee colony through the winter. A common rule of thumb is to leave at least 60 pounds (27 kilograms) of honey in the hive, though this can vary based on your local climate and the size of your bee colony.

After the harvest, return any wet frames to the hive. The bees will clean up the remaining honey and repair the comb, preparing it for the next season. This not only minimizes waste but also helps the bees maintain their hive.

In conclusion, harvesting honey is a delicate balance between taking what we need and ensuring the health and sustainability of the bee colony. With the right timing, tools, and techniques, you can enjoy the fruits of your labor while contributing to the well-being of these essential pollinators.

Common Problems and Solutions

Beekeeping on your backyard homestead can be a rewarding endeavor, offering the sweet reward of honey and the satisfaction of contributing to the health of the environment. However, like any agricultural pursuit, it comes with challenges. Understanding common problems and their solutions is crucial for maintaining a healthy bee colony and ensuring a successful harvest.

One of the most frequent issues beekeepers face is the Varroa mite infestation. These tiny parasites can wreak havoc on bee colonies by attaching themselves to bees and sucking their hemolymph, weakening and eventually killing the host bee. To combat Varroa mites, regular monitoring of the bee population is essential. This can be done through methods such as the powdered sugar roll or alcohol wash to estimate mite loads. If mite levels are high, treatments such as organic acids (formic or oxalic acid), essential oils (thymol-based products), or synthetic acaricides may be used, following the manufacturer's instructions closely to avoid harming the bee colony.

Another common problem is hive beetles, which thrive in a hive's warm, humid environment and can quickly become a nuisance by laying eggs in the comb and spoiling the honey. Maintaining hive hygiene is critical to controlling hive beetles. Regularly removing weak or dead colonies that attract beetles and using beetle traps can help keep their numbers in check.

Swarming is a natural part of bee behavior but can become problematic if not managed properly. It typically occurs when the colony grows too large for the hive, and a new queen is produced, leading half of the colony to leave in search of a new home. This can significantly reduce your hive's productivity. To prevent swarming, ensure your bees have enough space by adding new boxes or frames, particularly in the spring when the colony grows rapidly. Regularly checking for and removing queen cells can also discourage swarming.

Diseases such as American Foulbrood (AFB) and European Foulbrood (EFB) pose significant risks to bee colonies. These bacterial infections can spread quickly and are often fatal. The best defense against foulbrood is prevention through good apiary management practices. This includes purchasing bees from reputable sources, regularly inspecting colonies for signs of disease, and maintaining cleanliness in and around hives. If a colony is infected with AFB, it is often recommended to destroy the hive to prevent the spread of the disease, as AFB spores are highly resistant to disinfectants.

Lastly, pesticide exposure can severely impact bee health, leading to weakened or dead colonies. Advocating for responsible pesticide use in your community and planting bee-friendly flora can help mitigate these risks. Additionally, placing hives away from treated fields and notifying neighbors of your beekeeping can help protect your bees from accidental exposure.

By staying vigilant and proactive in managing these common problems, you can ensure the health and productivity of your bee colonies. Remember, successful beekeeping is about addressing issues as they arise and preventing them through sound management practices. This benefits your backyard homestead and supports the broader ecosystem by contributing to the health and diversity of local bee populations.

The Importance of Bees in Your Garden

Bees play a pivotal role in the health and productivity of your garden, acting as the linchpin in the pollination process vital for many plants' reproduction.

Pollination, the transfer of pollen from the male parts of a flower to the female parts, is necessary for fertilization and, subsequently, the production of fruits and seeds. Bees are among the most effective pollinators, not by chance but by nature. As they move from flower to flower in search of nectar, they inadvertently transfer pollen, facilitating the growth of many of the foods and flowers we enjoy.

The importance of bees in your garden extends beyond the mere act of pollination. Their activity helps increase the quality and quantity of your crops. Well-pollinated plants tend to produce larger, more uniform fruits and vegetables, which is a boon for any homesteader looking to maximize their garden's yield. Moreover, a diverse bee population encourages plant diversity, which can help make your garden more resilient to pests and diseases.

Incorporating beekeeping into your backyard homestead also contributes to the broader ecosystem. Like many other pollinators, bees are threatened by various environmental pressures, including habitat loss, pesticides, and climate change. By providing a haven for bees, you're enhancing your garden's productivity and contributing to the conservation of these essential creatures.

To make your garden inviting to bees, consider planting various flowering plants that bloom at different times throughout the year. This ensures a continuous food source for the bees, encouraging their presence in your garden. Additionally, providing water sources, avoiding pesticides, and allowing for natural areas within your garden can create a hospitable environment for bees.

Embracing beekeeping and understanding the symbiotic relationship between bees and gardens can transform your backyard homestead. It's a step towards sustainability, promoting a healthier ecosystem while reaping the tangible benefits of higher crop yields and a more vibrant garden. As we move forward, we'll explore how to get started with beekeeping, ensuring you have the knowledge and tools to integrate these vital pollinators into your homesteading journey successfully.

Chapter Summary

- Beekeeping begins with understanding bee behavior, which is crucial for safety and productivity. This focuses on the complex and organized nature of bees, including their social structure, roles, and communication methods.

- The queen bee's primary role is reproduction; worker bees perform tasks like foraging and maintaining the hive, and drones exist solely for mating.
- Bees communicate through pheromones and the "waggle dance" to share information about food sources and hive health.
- Defensive behaviors in bees include stinging to protect the hive, and beekeepers must approach with respect, using protective gear and proper handling techniques.
- Seasonal behaviors affect beekeeping practices, with bees being active in warmer months and focusing on survival in winter, influencing hive care across seasons.
- Challenges such as diseases, pests, and environmental stressors require beekeepers to be vigilant and responsive to maintain hive health.
- Setting up a hive involves selecting a location, choosing the type of hive, assembling it with care, and gently introducing bees, emphasizing creating a sustainable environment for the bees.
- Regular hive maintenance, understanding and addressing common problems like Varroa mites and hive beetles, and preparing for winter are essential for a healthy bee colony. These activities lead to successful honey harvesting and contribute to the ecosystem.

5

PRESERVING THE HARVEST

A rustic farmer's shack with an abundance of fresh fruits and vegetables in baskets.

Canning Basics

Canning is a time-honored method of preserving food from your backyard homestead, allowing you to enjoy the fruits of your labor throughout the year. This section will guide you through the basics of canning, ensuring you have the knowledge to safely and effectively preserve your harvest.

Understanding the two primary canning methods is crucial: water bath canning and pressure canning. Water bath canning suits high-acid foods such as fruits, tomatoes (with added acid), pickles, jellies, and jams. The acidity in these foods prevents the growth of bacteria, making the water bath method sufficient. On the other hand, low-acid foods like vegetables, meats, and poultry require pressure canning to reach the higher temperatures necessary to eliminate the risk of botulism.

Before you begin, it is essential to assemble the right equipment. You'll need a large pot with a lid and a rack for water bath canning to keep the jars off the bottom. Pressure canning requires a specialized pressure canner designed to reach higher temperatures than a regular pot. Regardless of the method, you'll also need canning jars with new lids and rings, a jar lifter, a funnel, and a bubble remover/headspace tool.

Preparing your produce is the next step. Start with fresh, high-quality fruits or vegetables. Wash them thoroughly and prepare them according to your recipe, whether peeling, chopping, or crushing. Pay close attention to the recipe's instructions on preparing and adding additional ingredients, such as sugar, salt, or vinegar, which can be crucial for preservation.

Filling your jars correctly is critical to successful canning. Use a funnel to pack your prepared food into the jars, leaving the appropriate headspace as indicated by your recipe. This space is necessary for the expansion of food as it heats and creates a vacuum seal as it cools. After filling, use a bubble remover tool to release any trapped air bubbles, then wipe the rims clean before placing the lids and screwing on the bands until they are fingertip tight.

Processing your jars is the final step. Place them in your canner with enough water to cover them completely, then bring the water to a boil for water bath canning or follow your pressure canner's instructions for that method. Processing times vary depending on the food and your elevation, so consult a reliable source for accurate times. Once processed, carefully remove the jars and let them cool undisturbed for 12 to 24 hours. You'll know your jars are sealed properly when the lids are concave and do not flex when pressed.

The last step in the process is labeling and storing your canned goods. Write the contents and the date on each jar, then store them in a cool, dark place. Properly canned foods can last up to a year, sometimes longer, though quality may diminish over time.

Canning is a rewarding way to preserve your harvest, providing homegrown goodness long after the growing season. You'll find it a valuable addition to your homesteading skills with a bit of practice.

Freezing and Drying

Mastering the arts of freezing and drying is akin to unlocking new levels of self-sufficiency and culinary creativity in the journey of transforming your backyard bounty into a pantry full of provisions. These preservation methods extend the shelf life of your harvest and ensure that summer flavors can be savored long after the season has passed.

Freezing is perhaps one of the most straightforward and accessible methods of food preservation. It works by slowing down the activity of enzymes and preventing the growth of bacteria, yeasts, and molds that cause food spoilage and decay. Almost every home has a freezer, making this method particularly convenient.

To begin, select fresh, ripe produce at its peak quality. Wash and prepare the fruits or vegetables by peeling, chopping, or blanching as required. Blanching – briefly boiling and then plunging the produce into ice water – is critical for most vegetables. It halts enzyme activity that can cause flavor, color, and texture loss, even in frozen storage.

After preparing your produce:

1. Pack it into freezer-safe containers or bags, removing as much air as possible to prevent freezer burn.
2. Label each package with the date and contents.
3. Remember, while freezing preserves texture, flavor, and nutritional value, it's best to use frozen fruits and vegetables within 8 to 12 months for optimal quality.

Drying, one of the oldest food preservation techniques involves removing moisture from food to inhibit the growth of microorganisms and enzymes. When done correctly, dried foods are lightweight, space-saving, and can last for months or years.

There are several methods to dry foods: air drying, oven drying, and using a food dehydrator. Each method has its advantages and is suited to different types of food. Herbs and leafy greens, for instance, dry well in the gentle breeze of an airy, shaded spot. However, fruits, vegetables, and meats often require the consistent heat of an oven or dehydrator to remove moisture effectively.

Before drying, prepare your produce by washing, slicing, and sometimes pretreating to preserve color and nutritional content. Pretreatment can include blanching or dipping in solutions like ascorbic acid. Spread the prepared items in a single layer on drying racks or trays, ensuring good air circulation around each piece.

Once dried, store your foods in airtight containers in a cool, dark place. Check periodically for moisture or mold growth, which can occur if foods are not sufficiently dried.

Both freezing and drying preserve the fruits of your labor and provide a canvas for culinary experimentation. From smoothies and soups made from frozen produce to the complex flavors of dried fruits and vegetables rehydrated in innovative dishes, these methods allow the backyard homesteader to enjoy the harvest in many ways throughout the year.

As we continue to explore the spectrum of preservation techniques, the journey from garden to table evolves, revealing not just the practicality of these methods but the profound satisfaction of sustaining oneself from the yield of one's backyard.

Fermenting and Pickling

Fermenting and pickling are ancient methods that have preserved the bounty of gardens long before refrigerators and freezers became household staples. These techniques extend the shelf life of your harvest, enhance nutritional value, and introduce a delightful array of flavors to your table.

Fermentation is a metabolic process that produces chemical changes in organic substrates through the action of enzymes. In simpler terms, it's the transformation of your cucumbers into crunchy pickles or cabbage into tangy sauerkraut. This magic is primarily performed by lactobacilli, beneficial bacteria that thrive in an anaerobic (oxygen-free) environment. They feed on the natural sugars in food, producing lactic acid as a natural preservative. The beauty of fermentation lies in its simplicity and the minimal equipment required. With just a clean jar, some salt, water, and fresh produce, you can start fermenting. The key is to ensure that your vegetables are fully submerged under the brine to prevent mold growth, keeping the anaerobic process intact.

While similar to fermenting, pickling takes a slightly different route. It involves immersing fruits or vegetables in a solution of vinegar or brine. This acidic environment kills bacteria, preventing spoilage. Pickling can be quick, needing only a few hours, or a longer fermentation-like process. The variety of spices and herbs you can add to your pickling solution is vast, allowing for endless creativity. From classic dill pickles to exotic spiced peaches, the flavors you can achieve are limited only by your imagination.

Both fermenting and pickling do more than preserve food; they transform it. Fermented foods are known for their probiotic qualities, contributing to a healthy gut microbiome. Meanwhile, pickled foods acquire a unique taste that can elevate a meal from good to gourmet.

Remember to start small as you embark on your fermenting and pickling journey. Experiment with different recipes and quantities until you find what works best for you and your family. Whether it's the tangy crunch of sauerkraut on a homemade burger or the zesty zing of pickled radishes in a salad, these preserved delights are sure to add a new dimension to your meals.

By mastering these age-old techniques, you'll preserve your harvest, embrace a sustainable lifestyle, reduce food waste, and ensure that your pantry is stocked with nutritious, flavorful options year-round. So, let's roll up our sleeves and transform the fruits of our labor into fermented and pickled masterpieces that our ancestors would be proud of.

Making Jams and Jellies

Transforming your garden's bounty into delicious jams and jellies is a rewarding endeavor and a delightful way to preserve the season's flavors. This guide will help you through the process of making jams and jellies, ensuring you can enjoy the fruits of your labor long after the harvest has ended.

Making jams and jellies might seem daunting initially, but it's pretty straightforward once you grasp the basics. Jams are made from crushed or chopped fruits cooked with sugar, while jellies are made from the fruit's juice, sugar, and often pectin, resulting in a clear, firm product. Achieving the perfect set and flavor requires a balance of fruit, sugar, and acid.

The journey to delicious jam or jelly starts with high-quality fruit. Ideally, you should use ripe fruit from your garden, ensuring it has the perfect combination of natural sugars and pectin. Pectin is essential for helping your jam or jelly set, and while some fruits naturally have high levels of pectin, such as apples and citrus fruits, others may need the addition of commercial pectin.

For jams, wash your fruit thoroughly, remove any stems or pits, and chop it into small pieces. For jellies, you'll need to extract the juice by chopping the fruit, cooking it until soft with a bit of water, and then straining it through a jelly bag or cheesecloth.

Next, combine your prepared fruit or juice with sugar in a large, heavy-bottomed pot. The sugar sweetens your preserve and helps it set by drawing out water from the fruit. Add lemon juice or another acid to balance the sweetness and aid the setting process. If using commercial pectin, follow the packet instructions for when to add it to your mixture.

Bring your mixture to a boil over medium-high heat, stirring frequently to prevent sticking. Once it reaches a rolling boil and starts to thicken, test if your jam or jelly is ready using the "wrinkle test" by placing a small amount on a chilled plate. If it wrinkles when you push it with your finger, it's done.

To preserve your jam or jelly:

1. Process it in sterilized jars.
2. Fill your jars, leaving a quarter-inch headspace, then wipe the rims clean and seal them with lids and bands.
3. Process the jars in a boiling water bath for the recommended time based on your altitude and jar size.
4. Once processed, remove the jars and let them cool undisturbed for 12-24 hours. The lids should be concave and not flex when pressed, indicating a proper seal.

Homemade jams and jellies make for a treat for your family and thoughtful gifts. Label your jars with the contents and the date, and store them in a cool, dark place. Properly canned, they can last for up to a year. Once opened, keep them refrigerated and enjoy them within a few weeks. By mastering the art of making jams and jellies, you'll preserve the flavors of your garden and create lasting memories and traditions. Whether spread on fresh bread or served as a sweet accompaniment to a cheese platter, your homemade preserves will surely delight.

Storing Vegetables and Fruits

After mastering the art of making jams and jellies, it's time to turn our attention to another crucial aspect of preserving the bounty of your backyard homestead: storing vegetables and fruits. This process ensures that your produce's freshness and nutritional value are maintained for as long as possible, allowing you to enjoy the fruits of your labor throughout the year.

The key to successful storage is understanding the specific needs of each type of produce. Different vegetables and fruits have varying temperature, humidity, and light exposure requirements. You can significantly extend your shelf life by creating the right environment for each type of produce.

A cool, dark, and moderately humid environment is ideal for most root vegetables, such as carrots, beets, and potatoes. These conditions mimic the natural underground habitat where these vegetables thrive. Storing them in a cellar, basement, or even a cool, dark cabinet can help preserve their freshness for several months. Removing any soil clinging to the vegetables is important, but avoid washing them until you're ready to use them, as moisture can promote decay.

Leafy greens, on the other hand, require a slightly different approach. These vegetables tend to wilt quickly if not stored properly. Wrapping them loosely in a damp cloth or paper towel and placing them in a plastic bag in the refrigerator can help maintain their moisture and crispness for a week or more.

Fruits generally prefer a cooler environment than vegetables. Apples, pears, and other tree fruits can be stored in a cool basement or garage where temperatures remain just above freezing. However, be mindful of ethylene gas, which many fruits emit as they ripen. This gas can accelerate the ripening process of nearby produce, so it's wise to store ethylene-producing fruits separately from those sensitive to the gas.

Berries and soft fruits present a unique challenge due to their delicate nature. These fruits are best consumed soon after harvest but can be stored in the refrigerator for a few days if necessary. To extend their shelf life, consider spreading them on a tray to avoid crushing and storing them in a single layer in a container.

In addition to these specific storage techniques, general principles apply to all produce. Always inspect your fruits and vegetables before storing them and remove any damaged or decaying items to prevent the spread of rot. Proper ventilation is also crucial to prevent the buildup of ethylene gas and moisture, which can lead to spoilage.

By employing these storage methods, you can enjoy the vibrant flavors and nutritional benefits of your garden's produce beyond the harvest season. Whether you're storing root vegetables in a cool basement, wrapping leafy greens in damp cloths, or carefully managing the storage of fruits, each technique plays a vital role in extending the life of your harvest.

Creating a Root Cellar

One of the most rewarding endeavors in backyard homesteading is the ability to preserve your harvest for the colder months. After exploring the various methods of storing vegetables and fruits, it becomes clear that creating a root cellar is an indispensable step for any serious gardener or homesteader. This traditional storage method not only extends the shelf life of your produce but also maintains its nutritional value, making it a cornerstone of self-sufficiency.

A root cellar is a natural underground storage area that leverages the earth's constant temperature to store fruits and vegetables. Its beauty lies in its simplicity and the minimal energy it requires to function. Unlike modern refrigeration, a root cellar uses the earth's natural coolness and humidity to preserve produce. This method is eco-friendly and cost-effective, making it an ideal choice for the environmentally conscious homesteader.

To create your root cellar, start by selecting an appropriate location. Ideally, this should be a spot where the ground naturally slopes away from the cellar entrance to prevent water from pooling around or entering the storage area. It's also crucial to consider the water table in your area; you want to ensure that your cellar is not prone to flooding during heavy rains.

The size of your root cellar will depend on your storage needs and the space available on your property. A small cellar might suffice for a family looking to store a modest harvest, while larger operations may require a more spacious design. Regardless of size, insulation is critical. Straw bales, earth mounds, or modern insulation materials can be used to maintain the cellar's temperature and humidity levels.

Ventilation is another critical aspect of root cellar design. Proper airflow prevents the buildup of ethylene gas, which can cause produce to spoil prematurely. A simple ventilation system with an intake and exhaust vent can help maintain the optimal environment for your stored goods.

When outfitting your root cellar's interior, simplicity and functionality should guide your choices. Shelves and bins made from wood or wire mesh materials allow for organized storage and adequate air circulation around the produce. It's also beneficial to group fruits and vegetables by their storage requirements; some produce emits more ethylene gas than others, and separating these can minimize spoilage.

Incorporating a root cellar into your backyard homestead is not only a nod to traditional food preservation methods but also a practical solution for extending your garden's bounty. With some planning and effort, you can create a sustainable storage space that keeps your harvest fresh and nutritious throughout the year. This endeavor enhances your self-sufficiency and deepens your connection to the land and the cycles of nature.

Chapter Summary

- Canning is a method to preserve food, with water bath canning for high-acid foods and pressure canning for low-acid foods.
- Essential canning equipment includes jars, lids, a canner, and tools like a jar lifter and funnel.

- Preparing produce involves washing and possibly adding ingredients like sugar or vinegar for preservation.
- Proper jar filling and sealing are crucial, with headspace for expansion and airtight sealing to prevent spoilage.
- Process jars in a canner, with processing times varying by food type and elevation.
- Label and store canned goods in a cool, dark place for up to a year.
- Freezing and drying are other preservation methods, with freezing halting enzyme activity and drying removing moisture to prevent microorganism growth.
- Fermenting and pickling transform food, extending shelf life and enhancing flavor. Fermenting relies on bacteria, and pickling uses vinegar or brine.

6

RENEWABLE ENERGY ON THE HOMESTEAD

Vibrant sunset colors over a farm with wind turbines.

Solar Power Basics

Harnessing the sun's power has become an increasingly popular and practical way to generate electricity, especially for those looking to reduce their carbon footprint and achieve a more sustainable lifestyle on their backyard homestead. Solar power, with its promise of clean, renewable energy, offers a beacon of hope and independence for homesteaders. In this section, we'll delve into the basics of solar power, covering how it works, the types of solar systems available, and some considerations for installation and maintenance.

At its core, solar power converts sunlight into electricity using photovoltaic (PV) panels. These panels are made of semiconductor materials, such as silicon, which absorb sunlight and release electrons, generating an electric current. This process, known as the photovoltaic effect, is the foundation of solar power technology.

Two primary types of solar power systems for the backyard homestead are grid-tied and off-grid. Grid-tied systems are connected to the local utility grid, allowing homeowners to feed excess electricity back into the grid through net metering. This can offset electricity costs and generate income. On the other hand, off-grid systems are entirely independent of the utility grid, relying on batteries to store the electricity generated by the solar panels for use when the sun isn't shining. Off-grid systems are ideal for remote locations where connecting to the grid is impractical or too expensive.

When considering solar power for your homestead, assessing your energy needs is essential. This involves calculating your household's average energy consumption and determining the size and number of solar panels required to meet this demand. Factors such as the orientation and angle of your roof, local climate conditions, and potential shading from trees or buildings can all impact the efficiency of your solar power system.

Solar panel installation typically requires professional help, although DIY kits are available for those with the necessary skills and confidence. Ensuring your installation complies with local building codes and regulations, including obtaining permits and having the system inspected upon completion, is crucial.

Maintenance of solar power systems is relatively minimal. It primarily involves keeping the panels clean and free of debris to ensure maximum efficiency. Regular checks to ensure all components are functioning correctly can help extend the life of your system.

Incorporating solar power into your backyard homestead can provide numerous benefits, including reducing your reliance on fossil fuels, lowering your electricity bills, and contributing to a healthier planet. With technological advancements and increasing affordability, solar power has become an accessible and viable option for homesteaders embracing renewable energy.

Wind Energy Fundamentals

Harnessing the power of the wind is an ancient practice, refined over centuries into a sophisticated method of generating renewable energy. For the modern backyard homesteader, wind energy presents an invaluable opportunity to complement solar power, ensuring a more consistent and diversified approach to self-sufficiency in energy production.

Wind energy operates on a simple principle: converting kinetic energy from wind into mechanical power or electricity. This is achieved through a wind turbine, which captures the wind's force using blades connected to a rotor. The rotor then spins a generator to create electricity. For homesteaders, small-scale wind turbines offer a feasible and efficient method to generate power, especially in favorable wind conditions.

Before considering installing a wind turbine, it's essential to understand the fundamentals of wind energy, including assessing your local wind resources. Wind speed and consistency vary greatly by location, influenced by local topography and obstacles such as buildings or trees. An ideal site for a wind turbine is typically an open, exposed area with high average wind speeds. Tools and resources are available to help homesteaders evaluate their wind resource potential, including wind maps and anemometers for measuring local wind speeds.

The size and type of wind turbine suitable for a homestead depend on the energy needs and the wind resource assessment. Small wind turbines can range from under 100 watts, suitable for charging batteries and powering small appliances, to systems capable of generating several kilowatts, enough to power an entire home. The height of the turbine's tower also plays a critical role in performance, as wind speeds increase with elevation above ground level.

Installing a wind turbine requires careful planning and consideration of local zoning laws, permits, and potential impacts on neighbors and the environment. It's also important to consider integrating wind energy with other renewable energy sources, such as solar panels, to create a hybrid system that can provide more consistent power under varying weather conditions.

Maintenance is another critical aspect of wind energy on the homestead. While modern wind turbines are designed for durability and long-term use, regular inspections and maintenance are necessary to ensure optimal performance and prevent potential failures. This includes checking the turbine's blades, bearings, and electrical connections for wear and damage.

In conclusion, wind energy offers a promising avenue for backyard homesteaders to achieve energy independence and sustainability. By understanding the fundamentals of wind energy, assessing local wind resources, and carefully planning the installation and maintenance of a wind turbine, homesteaders can harness the power of the wind to complement their renewable energy portfolio. This approach contributes to a more resilient and self-sufficient homestead and supports broader environmental sustainability goals by reducing reliance on fossil fuels and lowering carbon emissions.

Rainwater Harvesting

The sky above us is an invaluable resource that often goes overlooked. Not for solar power, which certainly has its place, but for the life-giving rain it provides. Rainwater harvesting is not just an ancient practice rekindled for modern times; it's a practical, sustainable solution for water needs that aligns perfectly with the ethos of renewable energy.

Rainwater harvesting involves collecting and storing rainwater from rooftops, greenhouses, or other surfaces to be used later for irrigation, livestock, or even household needs after proper treatment. This method is a cornerstone of self-sufficiency on the homestead, reducing dependence on municipal water supplies and minimizing the ecological footprint of your homestead.

The beauty of rainwater harvesting lies in its simplicity and the variety of methods available for implementation, ranging from basic rain barrels to more sophisticated systems with cisterns, filters, and pumps. The system choice depends on your water needs, the space available, and your budget.

Starting with the basics, a rain barrel can be easily connected to a downspout from your roof, capturing water that would otherwise be lost to runoff. This water can then be used to water your garden, reducing your water bill and providing your plants with chlorine-free water that they will thrive on. For those looking to take a step further, incorporating first-flush diverters and filters can improve the quality of the water collected, making it suitable for a broader range of uses.

Installing a large cistern can provide a substantial water reserve for the more ambitious. These systems can be designed to supply water for all your irrigation needs and, with proper filtration and purification, can even provide potable water for your household. The initial investment may be higher, but the long-term savings and resilience it adds to your homestead are invaluable.

Beyond the practical benefits, rainwater harvesting contributes to managing stormwater runoff, reducing erosion and the burden on local water treatment facilities. It's a way to directly engage with the natural water cycle, fostering a deeper connection with the environment and promoting sustainable living practices.

Implementing a rainwater harvesting system on your homestead is not just about saving money or being eco-friendly; it's a step towards self-reliance and resilience. It complements other renewable energy initiatives, such as wind and solar power, creating a holistic approach to sustainable living. With careful planning and consideration of local regulations, rainwater harvesting can be a rewarding addition to your backyard homestead, ensuring you make the most of every drop of rain falling from the sky.

Geothermal Options

Harnessing the Earth's natural heat through geothermal options presents an intriguing and increasingly accessible avenue for energy self-sufficiency on the backyard homestead. Geothermal energy, derived from the Earth's internal heat, offers a stable and continuous power source, distinguishing it from other renewable resources that may fluctuate with weather conditions.

At the core of geothermal energy utilization for the homesteader is the geothermal heat pump system, which exploits the ground's constant temperature beneath our feet. Even just a few feet below the surface, the Earth maintains a nearly constant temperature, ranging from 45°F (7°C) to 75°F (24°C), depending on the location. This consistency can be leveraged to both heat and cool homes and outbuildings in an incredibly efficient manner.

The basic mechanism involves circulating a fluid, typically water or a water-antifreeze mix, through a loop of underground pipes. During the winter, this fluid absorbs the Earth's natural warmth and carries it into the home, where a heat pump extracts and distributes the heat. In the summer, the process reverses, extracting heat from the building and dispersing it into the ground, thus cooling the interior.

Implementing a geothermal system on a homestead requires an initial investment and some land for the underground loop system. However, the long-term benefits can be substantial. These systems are known for their longevity, low maintenance, and the ability to reduce heating and cooling costs by up to 70%. Moreover, they significantly decrease a homestead's carbon footprint, aligning well with the principles of sustainable living.

For those considering geothermal options, conducting a thorough site assessment is essential. Factors such as soil composition, land area, and local climate will influence the system's design and efficiency. Professional installation is recommended, as the setup requires specialized knowledge and equipment. Additionally, some regions offer incentives or rebates for geothermal systems, making it a financially viable option for many.

In embracing geothermal energy, homesteaders tap into a renewable and efficient power source and contribute to a more sustainable and resilient energy system. This approach aligns seamlessly with the broader goals of homesteading, which emphasize harmony with nature, self-sufficiency, and reduced reliance on external resources.

As we transition from exploring the potential of geothermal energy, it becomes clear that the journey towards a sustainable homestead continues with the adoption of renewable energy sources. The next step involves a holistic approach to energy use, focusing on conservation techniques that ensure the efficient utilization of these renewable resources, thereby maximizing the benefits and minimizing the environmental impact.

Energy Conservation Techniques

Understanding and implementing energy conservation techniques is as crucial as harnessing renewable energy sources in the journey toward a sustainable and self-sufficient homestead. This section delves into practical strategies to significantly reduce energy consumption, lower utility bills, and minimize environmental impact. By adopting these methods, homesteaders can ensure a more efficient use of the renewable energy systems they have in place, such as those discussed in the context of geothermal options.

One of the foundational steps in energy conservation is conducting an energy audit of your homestead. This process involves assessing where and how energy is used and identifying areas where improvements can be made. Simple changes, such as sealing leaks around doors and windows, adding insulation, and upgrading to energy-efficient appliances, can profoundly impact reducing energy waste.

Lighting is another area where significant savings can be achieved. Switching to LED bulbs, which use at least 75% less energy and last 25 times longer than incandescent lighting, can dramatically decrease energy consumption. Additionally, maximizing natural light during the day can reduce the need for artificial lighting, further conserving energy.

Water heating is a significant energy expense in many homes. To conserve energy, consider lowering the water heater's temperature setting, insulating the water heater tank and the first six feet of hot and cold water pipes, and using water-saving fixtures to reduce hot water usage. Solar water heaters can effectively reduce reliance on traditional energy sources for those looking to invest in renewable energy.

Appliances and electronics also contribute significantly to a household's energy footprint. Using energy-efficient appliances, unplugging devices when not in use, and employing smart power strips can help manage and reduce energy consumption. When purchasing new appliances, look for the Energy Star label, which indicates the product meets energy efficiency guidelines set by the U.S. Environmental Protection Agency.

Heating and cooling systems are among the largest energy consumers in a home. Regular maintenance of these systems and smart thermostat use can optimize their efficiency. Using fans instead of air conditioning can provide comfort at a fraction of the energy cost during warmer months. Strategies such as layering clothing, using thermal curtains, and ensuring proper insulation can keep the home warm without excessive heating in cooler months.

Lastly, embracing a lifestyle that prioritizes energy conservation can lead to innovative solutions and habits that further reduce energy use. Simple practices like cooking with a pressure cooker or solar oven, drying clothes on a line, and choosing manual tools over electric ones for gardening and yard work conserve energy and enhance the homesteading experience.

By integrating these energy conservation techniques, homesteaders can make significant strides toward achieving a more sustainable and energy-efficient lifestyle. These practices complement the existing renewable energy systems and pave the way for a more resilient and self-sufficient homestead.

Chapter Summary

- Solar power converts sunlight into electricity using photovoltaic (PV) panels, offering a clean, renewable energy source for homesteaders with grid-tied and off-grid options.
- Assessing energy needs, roof orientation, local climate, and potential shading is crucial for efficient solar power system installation, which may require professional help and adherence to local regulations.
- Wind energy, generated through small-scale turbines, complements solar power for a diversified renewable energy approach, requiring site assessment for wind resource potential.
- Rainwater harvesting, a sustainable method for water needs, involves collecting rain from surfaces for later use, with systems ranging from simple barrels to sophisticated cisterns.
- Geothermal energy, utilizing the Earth's constant underground temperature, provides efficient heating and cooling through a geothermal heat pump system. It requires an initial investment but offers long-term benefits.
- Energy conservation techniques, including conducting energy audits, using LED bulbs, and optimizing heating and cooling systems, enhance the efficiency of renewable energy systems and reduce environmental impact.
- Integrating renewable energy into the homestead involves careful planning. Starting with small

projects and potentially expanding, consider solar panels, wind turbines, and micro-hydro power based on individual needs and site conditions.

- Community engagement and shared knowledge among homesteaders are valuable tools for navigating the complexities of renewable energy integration and offering support and innovative solutions for sustainable living.

7

WATER MANAGEMENT

A modern water tank nestled among lush greenery in a garden.

Irrigation Systems for the Homestead

In backyard homesteading, mastering the art of water management is akin to unlocking one of nature's most vital secrets. Implementing an efficient irrigation system is a cornerstone for sustainable agriculture among the myriad techniques at the homesteader's disposal. This section delves into the various irrigation systems that can be adapted to a homestead's unique needs and constraints, ensuring that every drop of water is utilized to its fullest potential.

At the heart of the homestead, irrigation delivers water that mimics the natural hydration plants receive in their native habitats. This involves considering the timing, quantity, and method of water delivery to optimize plant health and minimize waste. While effective for small gardens, traditional methods, such as hand watering with a hose or watering can, quickly become impractical as the homestead scale expands. This is where more sophisticated systems come into play.

Drip irrigation emerges as a frontrunner for its precision and efficiency. Delivering water directly to the base of each plant through a network of tubing, emitters, and connectors minimizes evaporation and runoff, making it ideal for arid climates and water-conservation areas. The system can be customized to suit the layout of any garden. A timer can automate the watering process, saving time and ensuring consistent moisture levels.

Another system worth considering is soaker hoses. These porous hoses allow water to seep slowly along their length, providing a gentle, even watering perfect for raised beds and row crops. They are simpler to install than drip systems and can be covered with mulch to reduce water loss through evaporation further.

Sprinkler systems might be the answer for homesteaders with larger plots or various crops. While water usage is less efficient than drip and soaker hose systems, they can cover large areas quickly and are adjustable to meet different watering needs. However, they are best used in the early morning or late evening to reduce water loss to evaporation during the day's heat.

Rainwater harvesting is another critical component of a homestead's irrigation strategy. By collecting runoff from roofs and storing it in barrels or tanks, homesteaders can create a sustainable water source that reduces dependence on municipal systems and wells. This harvested rainwater can then be integrated into the irrigation system, providing a free, eco-friendly solution to water the garden.

Lastly, creating swales or rain gardens can enhance the efficiency of any irrigation system. These landscape features are designed to capture and hold rainwater, allowing it to percolate slowly into the soil, recharging groundwater, and providing moisture to plants over time. This method conserves water, reduces erosion, and improves soil health.

In conclusion, selecting the right irrigation system for a homestead is a multifaceted decision that depends on the local climate, soil type, water availability, and the specific needs of the crops being grown. By combining one or more of these systems with sustainable practices like rainwater harvesting and creating water-conserving landscape features, homesteaders can ensure their gardens thrive while stewarding the precious water resource.

Greywater and Its Uses

One of the most innovative and sustainable practices in backyard homesteading is the utilization of greywater, which is gently used water from bathroom sinks, showers, tubs, and washing machines. This resource can be crucial in efficiently managing a homestead's water needs. Greywater, distinct from fresh water and black water (sewage), is less polluted than black water and can be reused for various purposes without extensive treatment. However, it's essential to handle greywater carefully due to its potential content of dirt, food, grease, hair, and household cleaning products to avoid health risks and environmental harm.

The benefits of reusing greywater are significant. It can drastically reduce the demand on the main water supply, conserving fresh water for drinking and cooking, and lead to substantial savings on water bills. Additionally, using greywater for irrigation can enhance soil fertility by adding nutrients and promoting plant growth.

Setting up a greywater system requires thoughtful planning and compliance with local regulations, which vary considerably. Simple systems might directly divert greywater to gardens or orchards using gravity, while more complex setups could include filters, pumps, and surge tanks for temporary storage. Regardless of the system's complexity, using biodegradable, non-toxic soaps and detergents is crucial to avoid harming plants and soil.

To safely use greywater, avoid contact with edible plant parts and focus on watering the roots. Rotating greywater use with fresh water helps prevent the soil's build-up of salts or potential contaminants. Regular monitoring of soil pH and salinity is also beneficial for adjusting greywater usage and maintaining soil health.

Incorporating greywater into a backyard homestead's water management strategy is an intelligent step toward sustainability. By understanding its uses, benefits, and necessary precautions, homesteaders can effectively leverage this resource, contributing to a more self-sufficient and environmentally friendly living space. Exploring additional water features and conservation methods will further enhance the ability to create resilient and productive homesteads.

Creating Ponds and Water Features

In backyard homesteading, the addition of ponds and water features not only enhances the aesthetic appeal of your space but also plays a crucial role in sustainable water management. This section delves into the practical steps and considerations for creating these water elements, ensuring they contribute positively to your homestead's ecosystem.

Creating a pond or water feature begins with careful planning. First, consider the location. If you plan to include aquatic plants, they should be placed in an area with adequate sunlight. However, too much direct sunlight can lead to excessive algae growth, so a balance is necessary. Additionally, think about the proximity to your home and garden. A pond near your garden can serve as a source of irrigation water, while one closer to your home can provide a tranquil view and attract wildlife.

The size and depth of your pond are also important considerations. A deeper pond can support a wider variety of aquatic life and is less prone to rapid temperature fluctuations, which can stress fish and plants. On the other hand, a shallow pond is easier to construct and maintain. The size depends mainly on your available space and the pond's purpose. Whether for keeping fish, attracting wildlife, or simply for the beauty of water lilies blooming, your goals will dictate the design.

When it comes to construction, there are several methods to

choose from. Preformed pond liners are popular for small to medium ponds, offering ease of installation and a predetermined shape. For larger or custom-designed ponds, flexible liners allow for more creativity in shape and depth but require more skill to install correctly. Regardless of the type, ensuring a secure and leak-proof liner is paramount to the success of your pond.

Water circulation is another critical aspect. A good filtration system and a pump will keep the water moving and oxygenated, creating a healthier environment for plants and fish. Additionally, incorporating a waterfall or fountain adds a captivating visual and auditory element and aids in aeration.

Finally, introducing plants and fish can transform your pond from a simple water feature into a thriving ecosystem. Aquatic plants play a vital role in oxygenating the water and providing habitat for wildlife. When selecting fish, consider species well-suited to your climate and the size of your pond. They can help control mosquito larvae and algae, contributing to the overall health of your pond.

In conclusion, creating ponds and water features in your backyard homestead requires thoughtful planning and execution. By considering location, size, construction method, water circulation, and the introduction of plants and fish, you can create a sustainable and beautiful water element that enhances your homestead's ecosystem. It will serve as a water source for irrigation and provide a habitat for wildlife, contribute to biodiversity, and offer a serene spot for relaxation and enjoyment.

Water Purification and Filtration

Eensuring the purity and quality of your water sources is as crucial as any other aspect of self-sufficiency. After exploring the creation of ponds and water features, it's natural to delve into the methods and importance of water purification and filtration. This process guarantees a healthier environment for your homestead and secures a sustainable water source for both household and agricultural needs.

Water purification and filtration can be approached through various methods, each tailored to your homestead's specific needs and resources. The simplest form of filtration is the construction of a biofilter, which utilizes natural materials such as gravel, sand, and charcoal to remove impurities from water. This method is particularly effective for smaller water features or as a preliminary step in a more complex purification system.

More advanced purification methods may be necessary for those relying on rainwater collection or natural water bodies as their primary water source. Boiling is the most straightforward technique to eliminate pathogens, but it's not always practical for large volumes of water. Solar water disinfection, or SODIS, offers a sustainable alternative, using the sun's ultraviolet rays to purify water in transparent containers over time. Though effective, this method requires ample sunlight and time, making it less reliable in certain climates or during emergencies.

Chemical purification, using chlorine or iodine, is another option for homesteaders. While effective in killing bacteria and viruses, these chemicals must be used cautiously, as they can pose health risks if not properly dosed. It's essential to follow guidelines carefully and consider using a secondary filtration method to remove residual chemicals before consumption.

For those seeking a more hands-off approach, commercially available water filters and purifiers offer convenience and efficiency. These systems range from simple pitcher filters to complex whole-house systems that meet various needs and budgets. When selecting a commercial system, consider factors such as the volume of water it can process, the specific contaminants it targets, and the maintenance it requires.

Regardless of the chosen method, regular water source testing is paramount. This ensures that your purification and filtration efforts are practical and that your water remains safe for consumption and use around the homestead. Water testing kits are readily available and provide a simple way to monitor the quality of your water over time.

In conclusion, integrating effective water purification and filtration systems into your backyard homestead is essential for maintaining a healthy, sustainable lifestyle. By understanding the options available and tailoring them to your specific needs, you can ensure a reliable supply of clean water for your family and your homestead's various needs. As we move forward, the principles of managing water runoff will further enhance our ability to sustainably manage our precious water resources, highlighting the interconnectedness of all water management aspects in backyard homesteading.

Managing Water Runoff

After exploring the intricacies of water purification and filtration, it's crucial to address another vital aspect of water stewardship: managing runoff. This section delves into practical strategies for controlling and utilizing runoff water, ensuring that every drop serves a purpose in your homestead.

Water runoff can lead to soil erosion, water wastage, and potential flooding if not appropriately managed. However, with thoughtful planning and implementation, runoff can be transformed from a potential problem into a valuable asset for your garden and homestead.

One effective method for managing water runoff is the creation of rain gardens. Rain gardens are strategically placed where runoff accumulates, acting as natural filtration systems. By planting native shrubs, perennials, and grasses, you can enhance your landscape's beauty while reducing runoff and improving water quality. These gardens absorb and filter runoff, preventing pollutants from reaching water bodies and recharging the groundwater.

Another technique involves the installation of rain barrels or cisterns to capture rainwater from rooftops. This collected water can later be used for irrigation, reducing your reliance on municipal water supplies and lowering your water bills. It's a simple yet effective way to harness runoff, turning a potential excess into a valuable resource for watering plants.

Swales and berms offer additional solutions for managing runoff. Swales, shallow trenches that follow the contour of the landscape, capture runoff and allow it to slowly infiltrate into the soil, hydrating plants along its path. Berms, raised soil areas, can be used with swales to direct water flow away from structures and toward more beneficial areas.

Another strategy to consider is incorporating permeable paving into walkways and driveways. Permeable materials allow water to seep through, reducing runoff and replenishing groundwater. This approach not only aids in water management but also enhances the aesthetic appeal of your homestead.

Finally, mulching plays a crucial role in managing water

runoff. Applying a generous layer of organic mulch around plants and over garden beds helps retain soil moisture, reduces evaporation, and prevents soil erosion. Mulch acts as a sponge, absorbing water and slowly releasing it into the soil, making it available for plant roots.

By implementing these strategies, you can effectively manage water runoff on your homestead, turning potential challenges into opportunities for sustainability and resilience. As we transition to the next section, we'll explore how to further optimize water usage by conserving water in the garden, ensuring that our homesteading practices are as efficient and environmentally friendly as possible.

Conserving Water in the Garden

Conserving water in the garden is not just a practice but a necessity in the pursuit of a sustainable and efficient backyard homestead. With the growing awareness of water scarcity and the importance of resource management, gardeners are turning towards innovative and traditional methods to ensure their gardens thrive without wasting precious water. This section delves into practical strategies that can be seamlessly integrated into your gardening routine, ensuring your green space is both lush and environmentally conscious.

Mulching is a gardener's best friend when it comes to water conservation. By applying a generous layer of organic mulch around your plants, you suppress weeds and significantly reduce water evaporation from the soil. Mulch acts as an insulating layer, keeping the soil cool and moist and encouraging root growth and water retention. Organic mulches, such as straw, wood chips, or leaf litter, break down over time and enrich the soil with nutrients.

Another effective strategy is rainwater harvesting. By collecting rainwater from rooftops and storing it in barrels or tanks, you have a ready supply of water that can be used during dry spells. This reduces your reliance on municipal water supplies and ensures your plants benefit from chemical-free water. Simple systems can be set up to direct rainwater from downspouts into storage containers, making this a practical and eco-friendly solution for water conservation.

Choosing the right plants is crucial for a water-efficient garden. Opt for native or drought-tolerant species that are well-adapted to your local climate and soil conditions. These plants require less water and are more resistant to pests and diseases, reducing the need for frequent watering and chemical interventions. Creating a garden that harmonizes with the local ecosystem conserves water and supports biodiversity.

Drip irrigation systems represent a significant advancement in efficient water use. Drip irrigation minimizes waste and ensures that water goes precisely where it's needed by delivering water directly to the base of each plant. This method effectively reduces evaporation and runoff, making it ideal for vegetable gardens, flower beds, and even potted plants. With the addition of a timer, the system can be automated, saving time and further optimizing water use.

Finally, understanding your garden's watering needs can lead to significant water savings. Overwatering is a common issue that wastes water and harms plant health. By learning to recognize the signs of water stress in plants and adjusting your watering schedule accordingly, you can ensure your garden receives just the right amount of water. Early morning or late evening watering reduces evaporation, and using a rain gauge can help you keep track of natural precipitation, adjusting your watering practices accordingly.

Incorporating these water conservation strategies into your gardening practices contributes to a more sustainable homestead and fosters a deeper connection with the natural world. By being mindful of our water use, we can create thriving gardens that are resilient, productive, and in harmony with the environment.

Chapter Summary

- Efficient water management is crucial for backyard homesteading, emphasizing sustainability and self-sufficiency.
- Greywater, wastewater from baths, sinks, and washing machines (excluding toilet waste), can be repurposed for irrigation and other non-potable uses.
- Using biodegradable and plant-friendly detergents is essential for greywater recycling to protect the homestead's ecosystem.
- Greywater systems can range from simple setups, like rerouting washing machine water to plants, to more complex systems involving filtration for broader use.
- Local regulations on greywater use vary, and it's important to ensure compliance and consider the suitability of plants for greywater irrigation.
- Creating ponds and water features enhances the aesthetic and sustainability of a homestead, supports

a variety of aquatic life, and provides irrigation sources.

- Water purification and filtration methods, including biofilters, boiling, solar disinfection, and commercial systems, are vital for maintaining water quality for household and agricultural use.
- Managing water runoff through rain gardens, rain barrels, swales, berms, permeable paving, and mulching helps prevent soil erosion and water wastage and supports water conservation in gardening.

8

LIVESTOCK MANAGEMENT

A man standing in a field surrounded by sheep under a cloudy sky.

Selecting Livestock for Your Homestead

Selecting livestock for your backyard homestead is an exciting and pivotal decision. It marks the beginning of a closer relationship with your food sources and a step towards sustainable living. However, it's not a decision to be taken lightly. The right choice can lead to a rewarding homesteading experience, while the wrong one can cause unnecessary challenges.

Firstly, consider the size of your land. Space is a premium resource in backyard homesteading, and each animal you introduce requires a portion of it, not just for roaming but for shelter, feeding areas, and waste management. For instance, chickens need less space than goats or sheep, making them a popular choice for smaller homesteads. On the other hand, if you have more space, you might consider larger livestock like cows or pigs.

Next, think about the purpose of raising livestock. Are you looking for animals that will provide meat, milk, eggs, or a combination of these? Chickens can offer both meat and eggs, making them an efficient choice for many homesteaders. Goats and cows can provide milk, with the former requiring less space and potentially more manageable for beginners. If meat production is your primary goal, consider the types of meat you and your family prefer and select species and breeds accordingly.

Another critical factor is the care and time you can dedicate to your livestock. All animals require daily attention, but some need more than others. Chickens, for example, are relatively low maintenance, needing regular feeding, egg collection, and coop cleaning. In contrast, dairy animals like goats and cows demand a more significant time investment due to milking routines, which can be twice daily.

Your decision should also be influenced by the local climate and the livestock's adaptability to your environment. Some animals fare better in colder climates, while others thrive in warmer conditions. Researching breeds that can comfortably adapt to your local weather patterns will ensure their health and productivity.

Lastly, consider the legalities and regulations in your area regarding keeping livestock. Zoning laws can vary significantly, with specific areas allowing a wide range of animals while others may have strict restrictions. It's essential to be informed and compliant to avoid any legal issues down the line.

Selecting suitable livestock is a foundational step in creating a thriving backyard homestead. By carefully considering your space, the purpose of raising animals, the time you can commit, the adaptability of different livestock to your climate, and local regulations, you can make informed decisions that align with your homesteading goals and lifestyle. With the right choices, the animals you bring into your homestead will contribute to your self-sufficiency and bring joy and fulfillment to your daily life.

As we move forward, our next focus will be understanding your chosen livestock's nutritional needs and feeding strategies, ensuring their health, productivity, and well-being in your care.

Feeding and Nutrition

After selecting the right livestock for your homestead, it is crucial to dive into the essentials of feeding and nutrition to ensure their health, productivity, and well-being. This section provides a comprehensive guide to understanding and implementing a balanced diet for your animals, tailored to their specific needs and the resources available on your homestead.

Firstly, it's essential to recognize that each type of livestock has unique dietary requirements. Chickens, for example, thrive on a diet rich in grains, vegetables, and protein, often found in commercial poultry feed supplemented with kitchen scraps and free-range insects. On the other hand, ruminants like goats and sheep require a diet high in fibrous plants like hay, with additional grains or commercial feed to meet their energy needs, especially during pregnancy or lactation.

Quality and quantity of feed play pivotal roles in the health and productivity of your animals. Overfeeding can lead to obesity and health issues while underfeeding can result in malnutrition and decreased productivity. Calculating the correct portions based on the animal's weight, age, and productivity level is essential, and adjusting as necessary to maintain optimal health.

Water, often overlooked, is a critical component of livestock nutrition. Clean, fresh water must always be available to prevent dehydration and support overall health. The amount of water needed can vary significantly depending on the species, the weather, and the animal's diet and stage of life.

Understanding the nutritional value of different forages is critical for those looking to sustain their livestock on pasture. Rotational grazing practices can help maintain the health of the pasture and ensure that animals have access to the most nutritious plants. Supplementing with hay, silage, or commercial feed may be necessary during winter or dry periods when pasture quality declines.

Homesteaders should also know the nutritional supplements and minerals their livestock requires. These can include salt blocks, calcium for laying hens, or mineral mixes for goats and sheep. Deficiencies in these areas can lead to health issues, so monitoring and supplementation according to the animals' needs are crucial.

Lastly, it's important to recognize signs of nutritional deficiencies or imbalances in your livestock, such as poor coat quality, weight loss, reduced productivity, or health issues. Regular observation and, if necessary, consultation with a veterinarian can help you adjust your feeding strategies to address any problems.

By understanding and implementing these feeding and nutrition principles, you can ensure the health and productivity of your livestock, contributing to the sustainability and success of your backyard homestead.

Health Care and Wellness

Ensuring the health and wellness of your livestock in backyard homesteading is as crucial as providing them with proper nutrition. This involves essential practices, preventive measures, and treatments that support effective livestock management, helping your animals survive and thrive.

Regular health assessments are vital in maintaining a healthy flock or herd, allowing you to catch potential health issues before they escalate. Observing your animals daily helps identify signs of distress, changes in eating habits, abnormal discharge, or sudden weight loss, with each species having specific health indicators.

Preventive care through vaccination and deworming is pivotal in livestock health. It protects against common diseases and prevents parasitic infestations that can lead to malnutrition, disease, and even death. Consulting with a local veterinarian to establish a vaccination schedule and adhering to a judicious deworming schedule are essential steps. It's also important to note that overusing dewormers can lead to resistance, emphasizing the need for veterinary advice.

The role of nutrition in health cannot be overstated, with nutritional deficiencies or imbalances leading to various health issues, from weakened immune systems to reproductive problems. Ensuring your livestock's diet meets their specific nutritional requirements is crucial.

Introducing new animals to your homestead poses a risk of introducing diseases to your existing livestock, making quarantine practices essential. A quarantine period of at least 30 days allows for observation of new animals for any signs of illness before integration.

Despite the best preventive measures, emergencies can happen, underlining the importance of having a first aid kit for your livestock and knowing the basics of animal first aid. This includes having items like wound disinfectant, bandages, and tools for hoof care and knowing when to seek veterinary help. Effective health management also includes meticulous record-keeping, tracking vaccinations, deworming, illnesses, treatments, and other health-related events, which aids in managing your livestock's current health and making future management decisions.

In conclusion, the well-being of your livestock reflects the care and attention you provide. By implementing routine health checks, adhering to a preventive care schedule, and being prepared for emergencies, you can ensure the health and wellness of your backyard homestead's animals. As you move forward, remember that healthy, well-cared-for animals are the foundation of a successful breeding program.

Breeding and Population Management

In backyard homesteading, the sustainability and productivity of your homestead significantly depend on how you manage the breeding and population of your livestock. This involves a deep understanding of breeding practices, genetic diversity, and population control to maintain a healthy and thriving livestock population. Each livestock species has its specific breeding cycle, which is crucial for homesteaders to grasp.

For example, chickens lay eggs almost daily that can be incubated to hatch new chicks, whereas goats and sheep typically breed in the fall, resulting in spring births. Recognizing these cycles is essential for planned breeding, ensuring you have the necessary resources and space for new additions.

Selective breeding is another critical aspect, allowing homesteaders to enhance the quality of their livestock by choosing animals with desirable traits such as good temperament, high productivity, and hardiness for breeding. However, it's vital to keep genetic diversity in mind to avoid inbreeding, which can lead to health issues and reduced vitality.

Managing the population effectively is also crucial. This involves decisions on which animals to keep, sell, or process for meat based on the homestead's capacity and the family's needs. This helps keep the population sustainable, ensuring adequate resources for each animal.

Accurate and detailed record-keeping plays an invaluable role in breeding and population management. Recording each animal's birth date, parentage, health history, breeding dates, and other relevant details is crucial for making informed decisions about breeding practices and managing the livestock's overall health and productivity.

Ethical considerations are central to breeding and population management, emphasizing the importance of treating all animals with respect and care. This includes providing proper nutrition, shelter, and medical care and making humane decisions about culling or processing animals for meat.

Effective breeding and population management form the backbone of a thriving backyard homestead. You can ensure a healthy, productive, and sustainable livestock population by understanding breeding cycles, engaging in selective breeding, managing population levels, maintaining detailed records, and adhering to ethical standards. This contributes to the self-sufficiency of your homestead and the welfare of the animals in your care.

Fencing and Housing

Ensuring the well-being and safety of your livestock is a multifaceted task that requires a deep understanding of their dietary and health needs, as well as providing them with appropriate fencing and housing. These aspects are critical for their protection, comfort, and maintaining order and cleanliness within your homestead.

Fencing serves as the initial safety barrier, fulfilling several roles: it secures your animals within your property, shields them from predators, and manages their grazing patterns to avoid the overuse of any particular area. The fencing required varies with the type of livestock; for instance, poultry needs chicken wire or poultry netting, whereas larger animals like goats or sheep might require sturdier materials such as wood or metal.

Electric fencing is another versatile option that can deter predators and prevent escapes, though it's crucial to introduce your animals to it cautiously to prevent stress and injury. Regular maintenance to ensure the fence's functionality and to clear any vegetation that might short-circuit the current is also essential.

The primary aim of providing housing for your livestock is to offer a safe and comfortable shelter that protects them from the elements and predators while also playing a significant role in health management. Adequate ventilation is key to avoiding respiratory problems, and surfaces that are easy to clean help maintain hygiene and minimize disease risks. The design of your animal housing should cater to the specific needs of your livestock; for example, laying hens requires nesting boxes, whereas goats need a dry, draft-free space.

It's essential that the size of your housing can accommodate the growth of your herd or flock, ensuring ample space for all animals. Incorporating natural light can enhance the well-being of your animals and assist in regulating their biological cycles, with windows or skylights being beneficial, provided they are secure and predator-proof.

The integration of fencing and housing into your homestead should be seamless, not only meeting the needs of your animals but also fitting well with your property's overall design and functionality. The placement of animal housing should be considered for its impact on daily routines, such as feeding and cleaning, emphasizing accessibility and convenience. The choice of materials and the design should also align with your homestead's aesthetic, with sustainable and locally sourced materials offering an environmentally friendly boost to your setup.

In essence, careful planning and consideration in the fencing and housing of your livestock are pivotal to the success and sustainability of your backyard homestead. By focusing on your animals' safety, comfort, and health, you foster a harmonious environment that supports their well-being and your homesteading aspirations.

Managing Manure

After ensuring your livestock have a secure and comfortable environment, the next logical step in managing a backyard homestead involves addressing the inevitable byproduct of animal husbandry: manure. This section delves into the practicalities of managing manure in a way that benefits your homestead, keeps your animals healthy, and minimizes environmental impact.

Firstly, it's essential to understand the value of manure. Far from mere waste, manure is a potent organic resource, rich in nutrients that can significantly enhance soil fertility and structure. However, it must be managed correctly to unlock its potential without causing harm.

Establishing a regular cleaning routine is the cornerstone of effective manure management. Daily removal of manure from animal enclosures prevents the buildup of harmful gases, reduces the risk of disease, and keeps your animals clean and comfortable. Equip yourself with the necessary tools—shovels, forks, and wheelbarrows—to make this task as efficient as possible.

Once collected, decide on a manure management strategy that aligns with your homestead's size and gardening needs. Composting is a highly recommended approach, transforming manure into a nutrient-rich, soil-enhancing material. A well-maintained compost pile requires balancing carbon-rich materials, like straw or leaves, with the nitrogen found in manure. This balance encourages aerobic decomposition, minimizing odors and killing off pathogens. Turn your compost regularly to aerate it; within a few months, you'll have a valuable addition to your garden beds.

Creating a manure spreader system might be viable for those with more land. This method directly applies aged or composted manure to fields as a natural fertilizer. However, it's crucial to understand local regulations regarding manure application to prevent nutrient runoff into waterways, which can cause environmental harm.

Consider partnering with local gardeners or community gardens in smaller homesteads, where composting might not be feasible due to space constraints. Many are eager for organic matter to enrich their soils and may be willing to collect and compost their manure.

Lastly, managing manure is not just about disposal or recycling; it's also about health and safety. Always wear protective gear when handling manure to avoid direct contact with pathogens. Ensure your storage or composting site is well away from water sources to prevent contamination.

By viewing manure management not as a chore but as an integral part of your homestead's ecosystem, you can create a sustainable cycle that benefits your soil, plants, and, ultimately, your table. With the right approach, what was once waste transforms into black gold, enriching your homestead and ensuring its productivity for years.

Chapter Summary

- Choosing suitable livestock involves considering land size, animal purpose, care needs, climate, and legalities for a successful homestead.
- Essential animal health and productivity hinge on tailored nutrition, including balanced diets, water, forage understanding, and mineral supplements.
- Livestock wellness relies on regular health checks, vaccinations, deworming, emergency plans, and quarantine for new arrivals.
- Sustainable homesteading demands knowledge of breeding cycles, selective breeding, population control, record-keeping, and ethical practices.
- Safety and comfort for livestock require proper fencing and well-designed shelters for protection and health management.
- Effective manure management, including regular cleaning, composting, and adherence to regulations, benefits soil fertility and environmental health.
- Selecting livestock necessitates evaluating available space, animal raising goals, care time, climate adaptability, and legal compliance.
- Sustainability and productivity in homesteading are achieved through comprehensive livestock management, including feeding, healthcare, and breeding.

9

HOMESTEAD CARPENTRY

A woman in a workshop measuring and sawing wood for a project.

Basic Carpentry Skills

Mastering basic carpentry skills is a cornerstone of self-sufficiency that empowers you to create and maintain your structures in your homestead. Whether you're a novice with a hammer or someone with some experience looking to refine your skills, understanding carpentry fundamentals will be the foundation for all your homestead projects, including constructing coops and pens for your animals.

Safety is paramount. Always wear appropriate safety gear, including gloves and eye protection, and ensure that your workspace is clean and free of hazards. Familiarize yourself with your tools before starting any project and understand their proper use and maintenance.

Measuring and marking accurately are the bedrock of successful carpentry. Invest in a quality tape measure, a carpenter's square, and a level. Learning to measure twice and cut once will save you time, materials, and frustration. Practice marking your materials clearly and precisely, as accurate cuts depend on accurate markings.

Next, cutting materials is a skill that requires patience and practice. Whether using a handsaw or a power saw, the key is making smooth, controlled cuts. Start with straight cuts before moving on to more complex angles. Remember, the type of saw blade you use can significantly affect the quality of your cut, so choose accordingly based on the material you're working with.

Joining materials together is where your projects start to take shape. Nailing, screwing, and gluing are basic techniques you'll use regularly. Understanding when and how to use each method is crucial. For instance, screws provide more strength and are easier to remove than nails, making them ideal for structures that may need disassembly or adjustment. Glue can add extra stability to joints but requires clamping and drying time.

Finally, finishing your projects improves their appearance and protects them from the elements. Sanding, painting, or staining your creations will extend their life and enhance the aesthetic appeal of your homestead.

As you hone these basic carpentry skills, you'll be better equipped to tackle various projects, from simple repairs to more complex constructions like animal coops and pens. Each project will build your confidence and expand your capabilities, bringing you closer to the self-sufficient homestead of your dreams.

Building Coops and Pens

After mastering basic carpentry skills, you're poised to tackle one of the most satisfying backyard homestead projects: constructing coops and pens. These structures are vital for your poultry and livestock's safety, comfort, and productivity. This guide will walk you through the entire process, from initial planning to construction, ensuring that even novices can succeed.

The first step, before even touching any tools, is thorough planning. You need to consider the size of your flock or herd and their specific requirements. Chickens, for instance, need nesting boxes and roosting bars, whereas goats require robust fencing to prevent them from escaping. It's essential to research the needs of the animals you plan to keep and design your structure to meet those needs.

Choosing the right materials follows planning. Treated lumber is often the go-to choice for many coops and pens because of its durability and resistance to decay. However, it's crucial to ensure that any materials you use are safe for animals and do not release harmful chemicals. Corrugated metal or shingles can offer sufficient protection against the weather for roofing. The essential tools for this project include a saw, hammer, drill, screws, and nails.

When it comes to construction, several tips can help ensure success. A solid foundation prevents predators from burrowing into your coop or pen. A buried hardware cloth or a concrete base can offer additional security. Proper ventilation is crucial for animal health, yet it's essential to safeguard against drafts. Placing vents or windows strategically can achieve this balance.

Design your structure with maintenance in mind; doors should be wide enough to facilitate cleaning, and nesting boxes should be easy to access for egg collection. Security against predators is paramount, so cover all openings with hardware cloth and secure doors with predator-proof latches. Remember your animals' comfort; insulation can help with temperature control, and perches and hideaways offer security.

Adding a few finishing touches can significantly enhance your project once the structure is built. Painting or staining the wood not only extends its lifespan but also improves the aesthetics of your homestead. Personal touches like decorative trim or a weathervane can add character to your project.

Constructing coops and pens is a rewarding endeavor that boosts the functionality and self-sufficiency of your backyard homestead. With meticulous planning, the appropriate materials, and hard work, you can create a secure and comfortable home for your animals that will last years.

Constructing Raised Beds and Trellises

Transitioning from the foundational structures for your animals, we delve into the critical components that will bolster your plant life in your backyard homestead: the construction of raised beds and trellises. These elements are functional, enhancing the productivity and health of your garden and adding an aesthetic appeal to your homestead. With a few tools, some lumber, and some effort, you can create durable and efficient structures that will benefit your garden for years.

Raised beds are a vital feature of efficient backyard gardening, offering improved soil conditions, better drainage, and easier weed and pest control, alongside making gardening spaces more accessible. To construct a raised bed, you'll need untreated lumber like cedar or redwood, which is naturally rot-resistant, galvanized screws or nails, a drill or hammer, landscape fabric, and a mix of soil and compost.

Start by deciding the dimensions of your bed. A standard size is 4 feet by 8 feet, allowing easy access from either side. Cut your lumber to size, assemble a simple box frame, and, if building taller beds, reinforce the corners to prevent bowing. Place the frame in its desired location, line it with landscape fabric, and fill it with soil and compost.

Building trellises is essential for supporting climbing plants such as beans, peas, and cucumbers. They maximize vertical space and can improve yields and air circulation around plants. You'll need wooden stakes or metal poles, wire mesh or netting, strings or garden twine, a hammer or mallet, and staples or ties.

Decide the location and size of your trellis, drive the stakes or poles into the ground, and if using wooden stakes, connect the tops with a horizontal beam for stability. Attach your climbing material securely, ensuring it's tight and stable. For strings or twine, a grid pattern or vertical lines will support the plants as they grow.

Both raised beds and trellises are customizable to fit the size and style of your garden, improving its functionality and beauty.

With these structures in place, your backyard homestead is well on its way to becoming a productive and sustainable oasis.

Remember, the longevity of these structures relies on regular maintenance and occasional repairs. This ensures they continue to effectively support your homesteading efforts and allow you to enjoy the fruits of your labor season after season.

Repair and Maintenance

In backyard homesteading, the ability to maintain and repair your carpentry projects is as crucial as constructing them. This section delves into the essential skills and knowledge every homesteader should possess to ensure their wooden structures remain sturdy, functional, and aesthetically pleasing over time.

Regular inspection is vital. Seasonal changes can cause wood to expand, contract, and sometimes warp. Periodically checking your raised beds, trellises, fences, and other wooden structures for signs of wear and tear can help you catch issues before they become significant problems. Look for loose screws or nails, cracks in the wood, or any signs of rot or pest infestation.

When it comes to repairs, having a basic toolkit is indispensable. This should include a hammer, screwdrivers, a saw, a drill, sandpaper, and a set of nails and screws. A wood filler can patch small holes or cracks for more specialized tasks. When larger sections of wood are damaged, they may need to be cut out and replaced entirely. This is where your saw and drill will come in handy, allowing you to remove the damaged section and secure a new piece.

Maintenance also involves protecting your wood from the elements. If your structures are made from something other than treated or naturally rot-resistant wood, consider applying a wood preservative. These products can significantly extend the life of your projects by protecting against decay, insects, and weathering. For an added layer of protection, especially for structures that come into direct contact with soil, a coat of paint or stain can beautify your project and seal the wood from moisture.

Another aspect of maintenance is ensuring the stability of your structures. Over time, the ground can shift, causing posts to lean or become unstable. Regularly check the alignment of posts and supports, and be prepared to reinforce them if necessary. This might involve digging around the base of a post to reset it to a more stable position or adding additional supports to a structure showing signs of strain.

Lastly, consider the lifecycle of your carpentry projects. Even with the best care, all wood eventually reaches the end of its useful life. Planning for this inevitability means designing your projects so they can be easily repaired, parts can be replaced, or the entire structure can be dismantled and repurposed or composted if made from untreated wood.

Incorporating these repair and maintenance practices into your routine ensures that your backyard homestead remains a place of beauty, productivity, and sustainability. This will save you time and money in the long run and deepen your connection to the land and the structures you build upon it. As we move forward, we'll explore more DIY projects that can enhance your homesteading experience, each requiring a blend of creativity, skill, and a willingness to learn and adapt.

DIY Homestead Projects

Embarking on DIY homestead projects is not just a way to save money; it's a journey into self-reliance and creativity. Whether crafting a cozy chicken coop, building sturdy raised garden beds, or assembling a functional compost bin, each project you undertake adds value and versatility to your homestead. This section will guide you through several foundational projects, offering practical advice to ensure your carpentry endeavors are both successful and satisfying.

Building a Chicken Coop

A chicken coop is essential for keeping your poultry safe and comfortable. Consider size (allowing 2-3 square feet per chicken inside the coop), ventilation, insulation, and predator-proofing when planning your coop. Use durable materials to construct the frame and walls, and ensure the coop is elevated off the ground to prevent moisture and predators from getting in. Adding nesting boxes and a roosting area will make it a welcoming home for your chickens.

Constructing Raised Garden Beds

Raised garden beds improve drainage and soil quality and reduce back strain when tending to your plants. To build one, you'll need untreated lumber, screws, and a good-quality soil mix. Assemble the frame to your desired size, ensuring it's not too wide that you can't reach the center from the sides. Fill it with a mix of compost, topsoil, and other amendments to create a fertile growing environment for your plants.

Assembling a Compost Bin

A compost bin is a simple yet effective way to recycle kitchen scraps and yard waste into nutrient-rich compost for your garden. You can build a basic bin using pallets or wire mesh. The key is ensuring proper aeration and easy access to turn the compost. Consider a three-bin system for a continuous supply: one bin for fresh scraps, one for compost in progress, and one for finished compost.

Crafting a Rainwater Harvesting System

Collecting rainwater is an eco-friendly way to water your garden. A simple rainwater harvesting system can be constructed using gutters, downspouts, and a storage container. Ensure the container is covered to prevent debris and mosquitoes from getting in. Also, consider installing a first flush diverter to improve water quality by diverting the initial runoff, which may contain contaminants from the roof.

Creating a Tool Shed

A well-organized tool shed can significantly enhance your efficiency and enjoyment of homestead projects. It doesn't have to be large; even a small space can be maximized with shelves, hooks, and bins. Use recycled materials where possible to keep costs down. Ensure your shed is weatherproof and secure to protect your tools from the elements and potential theft.

As you embark on these projects, remember that carpentry is as much about the process as it is about the outcome. Take your time, measure twice, cut once, and don't be afraid to make mistakes—they're often the best learning opportunities. With each project you complete, you'll improve your homestead and build a repertoire of skills that will serve you for years to come.

Safety and Tools

Transitioning from the realm of DIY homestead projects, where creativity and self-reliance are paramount, we now delve into the critical aspects of safety and the essential tools required for homestead carpentry. This section is designed to equip you with the knowledge and tools necessary to safely and effectively bring your carpentry projects to life, ensuring that your homestead not only thrives but does so with an unwavering commitment to safety.

Before we explore the tools that will transform raw materials into functional homestead assets, it's imperative to underscore the importance of safety. While carpentry is rewarding, it also comes with its share of risks, from minor cuts and bruises to more serious injuries. Therefore, adopting a safety-first mindset is non-negotiable.

- **Personal Protective Equipment (PPE):** Always wear appropriate PPE. This includes safety goggles to protect your eyes from flying debris, gloves to safeguard your hands from splinters and cuts, ear protection to shield your ears from the noise of power tools, and dust masks or respirators, especially when working with treated lumber or creating a lot of sawdust.
- **First-Aid Kit:** Keep a well-stocked first-aid kit readily accessible in your workshop. Accidents can

happen, and being prepared to treat minor injuries immediately is crucial.

- **Workshop Cleanliness and Organization:** A cluttered workspace is a hazardous workspace. Regularly clean your workshop to remove sawdust, which can be a fire hazard and a health risk. Ensure tools are stored correctly and ample space to work without tripping or knocking something over.

With safety protocols in place, let's focus on the tools that are the backbone of any homestead carpentry project. Whether building a chicken coop, repairing a fence, or creating custom furniture, having the right tools is essential.

- **Measuring and Marking Tools:** Precision is key in carpentry. Tools such as tape measures, rulers, squares, and marking gauges will ensure your projects are aesthetically pleasing and structurally sound.
- **Saws:** Depending on the project, a variety of saws will be required. Hand saws, circular saws, and jigsaws each serve different purposes, from rough cuts to delicate, detailed work.
- **Hammers and Mallets:** From driving nails to fitting joints, hammers and mallets are indispensable. A good claw hammer can serve multiple purposes, while a rubber mallet can be used for more delicate tasks that require a softer touch.
- **Chisels and Planes:** Chisels and planes allow for precision shaping and smoothing of wood for fine

woodworking projects. These tools require skill and practice but are essential for detailed work.

- **Power Drill and Bits:** A power drill, along with a variety of drill bits, is crucial for drilling holes for screws or dowels. It's also useful for pilot holes to prevent wood splitting.
- **Safety Equipment:** No tool is as important as your safety equipment. This includes items already discussed under the safety-first heading.
- **Maintenance:** Keeping your tools clean, sharp, and in good working order will make your projects more enjoyable and safer. Regular maintenance includes cleaning after each use, sharpening blades, and checking for wear or damage.

In conclusion, transitioning from planning and designing your DIY homestead projects to the actual building phase requires a shift in focus toward safety and tool mastery. By understanding and implementing safety protocols and equipping yourself with the essential tools of the trade, you're well on your way to successfully tackling a wide range of carpentry projects that will enhance your homestead's functionality and aesthetic appeal. Remember, each project completed safely and skillfully is a step towards a more self-sufficient and rewarding homestead life.

Chapter Summary

- Mastering basic carpentry skills is essential for backyard homesteading, enabling the construction of structures like animal coops and pens.
- Safety is paramount in carpentry; wear protective gear and ensure a clean workspace.
- Accurate measuring and marking are crucial for successful carpentry, as is practicing smooth, controlled cuts with appropriate tools.
- Joining materials through nailing, screwing, and gluing is fundamental, with each method serving different purposes in project construction.
- Finishing projects with sanding, painting, or staining protects them from elements and enhances their appearance.
- Building animal coops and pens requires planning, selecting safe materials, and constructing with security and comfort in mind.
- Constructing raised beds and trellises improves garden productivity and aesthetics, requiring basic tools and materials like untreated lumber and landscape fabric.
- Regular maintenance and repair of carpentry projects ensure their longevity and functionality. This involves inspections, the use of protective finishes, and structural stability.

10

NATURAL MEDICINE AND HOMESTEAD HEALTH

Freshly picked herbs and flowers on a garden table.

Growing Medicinal Herbs

Growing medicinal herbs in your backyard can be a fulfilling and practical addition to your homestead. It offers a natural approach to health and wellness while enhancing your garden's biodiversity.

Starting with the selection of the right herbs is crucial. Opt for those known for their healing properties and suitability to your climate, such as lavender for relaxation, chamomile for digestive health, and echinacea for immune support. Tailor your choices to meet your household's health needs for a more personalized garden.

Before planting, it's essential to assess your garden's conditions. Medicinal herbs generally need well-drained soil and a minimum of six hours of sunlight daily. Be cautious with herbs like mint, which can spread aggressively; growing them in containers can help keep them in check. Testing and amending your soil will give your plants the best start.

The planting process involves spacing your herbs appropriately to avoid overcrowding and ensuring they receive enough water, particularly in their early growth stages and during dry periods. However, be mindful not to overwater, as many medicinal herbs prefer drier conditions. Mulching can help maintain soil moisture and control weeds.

Harvesting your herbs at the right time is critical to maximizing their medicinal properties. This is typically before they flower when their oils and flavors are most concentrated. Harvest in the early morning for the best results. Drying the herbs by hanging them in small bundles in a warm, airy spot away from direct sunlight is an effective preservation method. Store the dried herbs in airtight containers in a dark place to maintain their potency.

As your experience with medicinal herbs grows, seek further knowledge through books, workshops, and herb societies to deepen your understanding of their uses and benefits. This ongoing learning will enhance your ability to use your garden's bounty for health and wellness.

Incorporating medicinal herbs into your homestead adds to its beauty and diversity and moves you towards a more self-sufficient approach to healthcare. With dedication and proper care, your medicinal herb garden will thrive, offering you a natural remedy resource right in your backyard.

Creating Herbal Remedies

Creating herbal remedies combines the joy of gardening with the science of natural healing, allowing you to transform your garden's bounty into potent natural medicines. The process begins with harvesting your herbs at the optimal time, usually in the morning when the dew has evaporated but before the sun gets too intense. This ensures the plants are at their peak in terms of active compounds. Flowers and leaves should be picked when the plants are in full bloom, while roots are best harvested in the fall when the plant's energy is concentrated below ground.

After harvesting, most herbs need to be dried in a well-ventilated, dark, and dry room. They should be spread out in a single layer and turned regularly to ensure even drying. Once they crumble easily between your fingers, they're ready to be stored in airtight containers away from direct sunlight to maintain their potency.

One of the simplest remedies is an herbal infusion, which involves pouring boiling water over dried or fresh herbs, covering them, and letting them steep. The steeping time varies depending on the part of the plant you're using but generally ranges from 10 to 20 minutes. These infusions can serve various purposes, from soothing teas to bases for creams and ointments.

Tinctures are another way to harness the power of your herbs. They create concentrated extracts by soaking the herbs in alcohol or vinegar. This method pulls the active compounds out of the herbs, resulting in a potent remedy after four to six weeks of shaking the mixture daily and then straining it.

Salves are particularly useful for topical applications, such as treating skin conditions or muscle pain. They're made by infusing herbs in a carrier oil and mixing them with beeswax to create a spreadable salve that can be applied directly to the skin.

Engaging in the creation of herbal remedies enhances your self-sufficiency and deepens your connection to the natural world and its healing powers. The success of these remedies lies in the quality of the herbs, and the care put into crafting them, offering a satisfying way to take control of your healthcare and well-being.

Natural First Aid

Having a well-prepared natural first aid kit is essential in backyard homesteading, where self-sufficiency and natural living are paramount. This section delves into the essentials of creating and utilizing a natural first aid kit, ensuring you're prepared for minor injuries and ailments that might occur on your homestead.

A natural first aid kit is not just a collection of items; it's a testament to the homesteader's ability to harness nature's bounty for health and healing. It complements the knowledge you've gained from creating herbal remedies, extending the use of those remedies into immediate and practical applications.

Herbal Salves and Balms are indispensable for treating cuts, scrapes, and bruises. Calendula salve, for example, is renowned for its healing and antiseptic properties. A range of salves can address different types of skin issues, from soothing insect bites to healing minor wounds.

- **Essential Oils:** A select few essential oils can be incredibly versatile in a first aid setting. Lavender oil, known for its calming and antibacterial properties, can be applied to burns and stings. Tea tree oil is another powerhouse, effective against fungal infections and as a disinfectant for cuts and scrapes.
- **Bandages and Wraps:** While not directly from the garden, having a supply of natural fiber bandages and wraps can help secure dressings made from your herbal preparations. These are essential for managing wounds and preventing infection.
- **Activated Charcoal:** This a must-have for any natural first aid kit. It effectively treats poisonings and stings, absorbing toxins from the body. It can also be used to make a poultice to draw out infections.
- **Herbal Tinctures and Extracts:** These concentrated herbal preparations are potent and have a long shelf life. Tinctures such as echinacea can support the immune system in the event of a wound, while others like witch hazel are invaluable for their astringent properties and help treat hemorrhoids and varicose veins.

- **Aloe Vera:** A fresh aloe vera plant or a bottle of pure aloe vera gel is essential for treating burns, sunburns, and skin irritations. Its soothing and healing properties make it a go-to remedy for skin issues.

Having a natural first aid kit is one thing; knowing how to use it effectively is another. It's crucial to familiarize yourself with each component of your kit—understanding their uses and contraindications. Remember, while natural remedies are effective, they are not a substitute for professional medical treatment in severe cases. Always assess the situation carefully and seek medical help when necessary.

Moreover, consider taking a course in natural first aid to enhance your skills and knowledge. This can empower you to act confidently and correctly when using your natural first aid kit.

A natural first aid kit is a vital component of the backyard homestead, bridging the gap between everyday accidents and the healing power of nature. By carefully selecting and understanding the uses of various natural remedies, you can ensure the health and well-being of your family and yourself. This approach to first aid complements the homesteading lifestyle and reinforces the importance of living in harmony with nature, prepared for whatever challenges come your way.

Preventative Health Practices

Adopting preventative health practices is not just wise—it's essential. This approach to wellness emphasizes the importance of maintaining balance and harmony within our bodies and our environment, leveraging nature's bounty to foster health and prevent illness.

Cultivating a diverse and nutritious diet is at the heart of preventative health on the homestead. The food we grow and consume is pivotal in our overall health. By focusing on various fruits, vegetables, herbs, and medicinal plants, homesteaders can get a wide range of nutrients, antioxidants, and phytochemicals that support the body's natural defenses. For instance, incorporating leafy greens rich in vitamins A, C, and K can bolster the immune system, while herbs like garlic and ginger are renowned for their antimicrobial properties.

Water is another cornerstone of good health, and ensuring access to clean, safe water is a priority for any homesteader. Rainwater harvesting systems and well water testing are practical measures to secure an adequate supply of water for both consumption and cultivation purposes.

Physical activity, an inherent part of homesteading life, is crucial in preventative health. The daily tasks of planting, harvesting, and tending to animals provide a source of exercise and foster a deep connection to the land and a sense of accomplishment. This physical engagement is complemented by the mental and emotional benefits of being in nature, which include reduced stress and improved mood.

Rest and relaxation are equally important. Homesteading can be demanding, and allowing time for restorative practices such as meditation, yoga, or simply enjoying the tranquility of nature is vital for maintaining balance. Sleep, too, is a critical component of health, and establishing a routine that promotes restful sleep is beneficial for both mental and physical well-being.

Preventative health on the homestead also involves being

proactive about potential health risks. This includes natural pest control methods to reduce exposure to harmful chemicals, proper handling, food storage to prevent foodborne illnesses, and personal protective equipment when necessary to prevent injuries.

Incorporating natural remedies and traditional healing practices into daily life complements these preventative measures. Many homesteaders find value in learning about the medicinal properties of plants they can grow, using these natural remedies to treat minor ailments and support overall health.

By embracing these preventative health practices, homesteaders can create a living environment that nurtures well-being, resilience, and a deep connection to the natural world. This holistic approach to health benefits the individual and their family and contributes to the homestead's sustainability and vitality.

Integrating Natural Medicine into Daily Life

In the rhythm of homestead life, integrating natural medicine into daily routines emerges as a seamless and intuitive process. This integration not only enhances homesteaders' overall health and resilience but also deepens their connection to the land and its cycles.

At the heart of this approach is the cultivation of a medicinal garden, a dedicated space where herbs and plants known for their healing properties thrive. Each plant is selected for its specific benefits, from the calming chamomile and lavender to the immune-boosting echinacea and elderberry. Homesteaders learn to tend these gardens with care, understanding that the health of these plants directly influences the potency of their healing properties.

Homesteaders harvest these plants at their peak and transform them into various natural remedies. Tinctures, salves, teas, and poultices become part of the homestead's toolkit for addressing minor ailments and injuries. For instance, a calendula salve can soothe skin irritations, while a peppermint tea eases digestive discomfort. These remedies, rooted in generations of traditional knowledge, offer effective and gentle alternatives to over-the-counter medications.

Beyond the medicinal garden, integrating natural medicine into daily life involves a holistic approach to food. The homestead kitchen becomes a place of alchemy where every meal is an opportunity to nourish and heal. Foods rich in vitamins, minerals, and antioxidants support the body's natural defenses. Fermented foods, such as sauerkraut and kombucha, introduce beneficial probiotics that promote gut health. By choosing seasonal and locally sourced ingredients, homesteaders ensure their diet is nutritious and sustainable.

Mindfulness practices also play a crucial role in this integrated approach to health. Homesteaders recognize the importance of tuning into their bodies and the natural world. Walking barefoot in the garden, practicing yoga at sunrise, or sitting quietly under a tree become vital components of daily life. These moments of connection and reflection are restorative for the mind and spirit and reinforce the homestead's rhythms and cycles.

In embracing these practices, homesteaders find that natural medicine is not just about treating illness but about nurturing a way of life in harmony with nature. It's a path that requires patience, observation, and a willingness to learn from successes and setbacks. Yet, the rewards are manifold, offering a sense of empowerment, well-being, and a deepened respect for the natural world.

As we move forward, it becomes clear that maintaining health on the homestead extends beyond individual practices to encompass the environment in which we live. The principles of cleanliness and order, vital to preventing disease and promoting well-being, are reflected in our living spaces' thoughtful organization and care. This holistic view of health, where personal practices and environmental stewardship are intertwined, sets the stage for exploring the next crucial aspect of homestead health: hygiene practices.

Homestead Hygiene Practices

In backyard homesteading, maintaining a high standard of hygiene is not just a matter of personal health; it's a cornerstone for sustainable living and the well-being of both the homestead and its inhabitants. As we delve into the practices underpinning homestead hygiene, we must recognize that these routines are deeply intertwined with our environment's natural cycles and resources.

First and foremost, water plays a pivotal role in homestead hygiene. Access to safe drinking, cooking, and cleaning water is paramount. Rainwater harvesting systems can be a sustainable water source, but it's crucial to implement proper filtration and purification methods to make this water safe for use. Additionally, greywater systems can recycle water from baths, sinks, and washing machines for irrigation, reducing waste and conserving resources.

Personal hygiene on the homestead goes beyond regular

bathing and handwashing. Natural soaps and shampoos, which can be made from ingredients grown right in your backyard, such as herbs and essential oils, offer a sustainable alternative to store-bought products. These homemade products reduce exposure to synthetic chemicals and minimize plastic waste and the carbon footprint of purchasing and transporting commercial hygiene products.

When maintaining a clean living environment, natural cleaning solutions like vinegar, baking soda, and lemon juice are effective, eco-friendly alternatives to harsh chemical cleaners. These ingredients can tackle many cleaning tasks, from disinfecting surfaces to removing stains without introducing harmful substances into your home or the environment.

Composting is a fundamental practice for any homestead in managing waste. By composting organic waste, you reduce the amount of trash sent to landfills and create a valuable resource for enriching the soil in your garden. Composting systems can vary from simple backyard piles to more sophisticated composting toilets, which offer a solution for recycling human waste into safe, usable compost.

Pest control is another critical aspect of homestead hygiene. Natural and preventative measures, such as companion planting, natural predators, and physical barriers, can effectively manage pests without using chemical pesticides. These practices protect your garden and livestock from pests and preserve your homestead's biodiversity and ecological balance.

Lastly, animal husbandry practices must prioritize cleanliness and disease prevention. Regular cleaning of animal living spaces, proper disposal of manure, and ensuring access to clean water and healthy food are essential measures for preventing illness and maintaining the health of your livestock.

By adopting these homestead hygiene practices, you not only safeguard the health of your family and animals but also contribute to the sustainability and resilience of your homestead. These practices demonstrate a commitment to living harmoniously with nature, leveraging its cycles and resources to nurture our health and the environment.

Chapter Summary

- Growing medicinal herbs on your homestead can enhance biodiversity and offer a natural approach to health and wellness. You can select herbs based on their healing properties and climate suitability.
- Essential steps in cultivating a medicinal herb garden include understanding your growing conditions, proper planting and care, and techniques for harvesting and preserving herbs.
- Expanding knowledge of medicinal herbs through reputable sources can improve the effectiveness and safety of using these plants for health benefits.
- Creating herbal remedies involves harvesting herbs at the right time, drying and storing them properly, and crafting infusions, tinctures, and salves to support health and well-being.
- A well-prepared natural first aid kit, including herbal salves, essential oils, and other natural remedies, is crucial for addressing minor injuries and ailments on the homestead.

- Adopting preventative health practices, such as a diverse diet, clean water, physical activity, and rest, along with natural pest control and hygiene practices, supports overall health and prevents illness.
- Integrating natural medicine into daily life through a medicinal garden, holistic food choices, mindfulness practices, and environmental stewardship enhances the health and resilience of homesteaders.
- Maintaining high hygiene standards on the homestead, including clean water access, natural personal care products, eco-friendly cleaning solutions, composting, pest control, and animal husbandry, is essential for sustainable living and well-being.

11

SUSTAINABLE LIVING

Sustainable urban landscape with rooftop solar panels.

Reducing Waste

Reducing waste is a pivotal chapter in the journey towards a more sustainable backyard homestead. It's not just about minimizing what we throw away but transforming our perspective on resources, seeing potential where we once saw refuse. This transformation begins with understanding the core principles of waste reduction: reduce, reuse, recycle, and rot (compost). These principles can be applied innovatively to support a thriving homestead while minimizing our environmental footprint.

Firstly, reducing waste means carefully considering our purchases and opting for items with minimal packaging or those made from sustainable materials. It's about making conscious choices, such as selecting bulk seeds or loose-leaf teas over individually packaged products. This mindset extends to every corner of the homestead, from the kitchen to the garden, encouraging us to question the necessity and longevity of each item we bring into our space.

Reusing is the next step in our waste reduction journey. It's a creative challenge that invites us to see the potential in items that might otherwise be discarded. Glass jars from kitchen staples can be repurposed for storing seeds, homemade jams, or as vessels for propagating plants. Old t-shirts become rags for cleaning or material for crochet projects. Even broken tools can offer parts for repair or new creations. This approach reduces waste and fosters a culture of resourcefulness and innovation.

Recycling is the most recognized aspect of waste management, yet it requires a thoughtful approach to be truly effective. Not all materials are equally recyclable, and understanding the specifics of local recycling programs is crucial. Homesteaders can prioritize easily recyclable materials and educate themselves on how to properly prepare items for recycling, ensuring they don't contaminate the stream. This awareness helps make recycling a more efficient and impactful practice.

Lastly, composting, or 'rot,' is a cornerstone of waste reduction on the homestead. We close the loop by transforming kitchen scraps, yard waste, and even certain paper products into rich compost, returning nutrients to the soil to support the next growth cycle. Composting reduces the amount of waste sent to landfills and enhances the health of our gardens, creating a direct link between our waste reduction efforts and our food production.

Each of these principles offers a pathway to reducing waste on the homestead, but they also represent a shift in mindset. It's about seeing the value in our resources and understanding the impact of our choices. By embracing these practices, we move towards a more sustainable homestead and contribute to a larger culture of conservation and respect for the natural world.

Eco-Friendly Homestead Products

Our choices about the products we use in our backyard homesteads play a pivotal role in the journey towards a more sustainable and self-sufficient lifestyle. Embracing eco-friendly homestead products is not just about reducing our environmental footprint; it's about nurturing a healthier ecosystem in our backyards and fostering a deeper connection with the natural world.

One of the first steps in this direction is to consider the tools and materials we use in our gardens and for animal care. Opting for tools made from sustainable materials, such as bamboo, wood, or recycled metal, can significantly reduce the demand for plastic products. These materials are not only more environmentally friendly but often offer a longer lifespan and a touch of natural beauty to your homestead.

Regarding pest control, natural and organic methods are at the heart of eco-friendly homesteading. Chemical pesticides, while effective, can harm beneficial insects, soil health, and even the water supply. Instead, consider introducing beneficial insects that naturally control pest populations, such as ladybugs and lacewings. Companion planting is another powerful technique, where certain plant combinations naturally repel pests or attract their natural predators.

In the realm of animal care, sustainable practices are equally important. Feeding your livestock and poultry with organic feed, free from genetically modified organisms (GMOs), not only supports their health but also ensures that your homestead remains a bastion of natural integrity. Moreover, consider implementing systems that allow your animals to contribute to the homestead's ecosystem, such as using chicken manure as a potent organic fertilizer.

Water conservation is another critical aspect of eco-friendly homesteading. Simple practices such as collecting rainwater in irrigation barrels can significantly reduce water usage. Drip irrigation systems, which deliver water directly to the base of plants, are an efficient way to minimize water waste in the garden.

Lastly, our cleaning and maintenance products around the homestead should be eco-friendly. Natural cleaning products, made from ingredients like vinegar, baking soda, and essential oils, are effective and safe for the environment and your family. For maintenance tasks, look for products with minimal packaging and those made from natural or recycled materials.

By integrating these eco-friendly products and practices into our homesteads, we contribute to a more sustainable world and enjoy the benefits of a healthier, more self-sufficient lifestyle.

As we move forward, sustainability principles can guide us in exploring even more ways to harness energy efficiently and sustainably, ensuring that our homesteads thrive for future generations.

Conservation Efforts

Conservation of natural resources plays a pivotal role in the journey toward a more sustainable lifestyle, especially within the confines of a backyard homestead. This section delves into practical and innovative strategies that homesteaders can employ to conserve water, soil, and biodiversity, thereby ensuring the health and productivity of their land for generations to come.

Water is a precious commodity in any garden or homestead. One of the most effective ways to conserve water is through rainwater harvesting. By installing rain barrels or designing a more complex rainwater catchment system, homesteaders can collect and store rainwater for irrigation purposes. This reduces reliance on municipal water supplies and decreases the energy footprint associated with water treatment and distribution. Additionally, employing drip irrigation systems or soaker hoses can significantly reduce water wastage by delivering water directly to the plant roots, where it's most needed.

Soil conservation is another critical aspect of sustainable homesteading. The health of the soil directly influences the health of the plants it sustains. Practices such as cover cropping, crop rotation, and applying organic mulches can protect soil from erosion, enhance its fertility, and maintain its moisture content. Cover crops, for instance, prevent soil erosion and fix nitrogen in the soil, reducing the need for synthetic fertilizers.

Similarly, incorporating compost into the garden beds improves soil structure and provides a rich source of plant nutrients, promoting a more vibrant and resilient ecosystem.

Biodiversity conservation is equally important in a backyard homestead. A diverse ecosystem is more resilient and productive, offering a natural defense against pests and diseases while supporting many pollinators and beneficial insects. Creating habitats for wildlife, such as birdhouses, bee hotels, and butterfly gardens, can enhance biodiversity and contribute to the ecological health of the homestead. Moreover, planting various crops, including native plants, can support local wildlife and promote a balanced ecosystem.

In conclusion, conservation efforts in a backyard homestead are not just about sustaining the land and its resources; they're about creating a harmonious and self-sustaining ecosystem that thrives on the principles of sustainability and resilience. By adopting water and soil conservation practices and promoting biodiversity, homesteaders can play a crucial role in preserving the environment for future generations while enjoying the bounty and beauty of their land today. When combined with sustainable energy practices, these efforts lay the groundwork for a holistic approach to sustainable living, setting the stage for community involvement and education in the broader quest for environmental stewardship.

Community Involvement and Education

The role of community involvement and education in the journey toward sustainable living cannot be overstated. While individual efforts in conservation and sustainable practices are crucial, collective action and shared knowledge within a community can significantly amplify the impact. This section delves into how backyard homesteaders can engage with their communities to foster a culture of sustainability and self-reliance.

One of the most powerful steps a backyard homesteader can take is to become an advocate for sustainability within their local community. This can be achieved through various means, such as organizing workshops, participating in local farmers' markets, or starting a community garden. Workshops can cover various topics, from organic gardening and composting to rainwater harvesting and renewable energy solutions. These educational initiatives spread valuable knowledge and create a platform for exchanging ideas and experiences, fostering a sense of community and collective learning.

Participating in local farmers' markets is another effective way to engage with the community. By selling or donating produce grown on your homestead, you provide others with access to fresh, locally-grown food and raise awareness about the benefits of sustainable farming practices. It's an opportunity to demonstrate the viability of backyard homesteading and inspire others to consider similar practices.

Starting or joining a community garden is yet another avenue for involvement. Community gardens can serve as living classrooms, offering hands-on experience in gardening and sustainable living to people of all ages. They can also help to address food insecurity in the community by providing fresh produce to those in need. Moreover, community gardens can become hubs for social interaction, strengthening community bonds and fostering a shared sense of responsibility for the local environment.

Education plays a pivotal role in community involvement.

By sharing knowledge and skills related to sustainable living, backyard homesteaders can empower others to make informed decisions about their lifestyle and consumption habits. This can be as simple as starting a blog or a YouTube channel focused on sustainable living practices or offering to speak at local schools and community centers. The goal is to ignite curiosity and passion for sustainable living in others, encouraging them to explore how they can contribute to a more sustainable future.

In conclusion, community involvement and education are essential components of sustainable living. By engaging with their communities, backyard homesteaders can extend the reach of their sustainability efforts, creating a ripple effect that encourages the widespread adoption of sustainable practices. Through workshops, participation in local markets, community gardens, and educational initiatives, homesteaders can play a crucial role in fostering a culture of sustainability and self-reliance. As we move forward, collective action and shared knowledge of communities will be key in navigating sustainable living challenges and ensuring a healthier, more sustainable future for all.

Living Off the Grid

Transitioning from a lifestyle that emphasizes community involvement and education, we delve into the essence of sustainable living through the lens of living off the grid. This approach embodies the principles of self-sufficiency and environmental stewardship and represents a profound commitment to reducing one's carbon footprint and living in harmony with nature.

Living off the grid in a backyard homestead involves a

series of strategic and thoughtful adaptations aimed at creating a lifestyle that is self-sustaining and minimally dependent on external utilities. This encompasses energy production, water sourcing, waste management, and food production, each of which contributes to a holistic approach to sustainable living.

Energy production is often the first consideration for those looking to live off the grid. Solar panels and wind turbines can be integrated into the homestead to harness natural resources for electricity. This reduces reliance on fossil fuels and ensures that the homestead remains powered during outages or disruptions in the grid. Additionally, using energy-efficient appliances and lighting and thoughtful design choices that maximize natural light and insulation can significantly reduce energy consumption.

Water sourcing is another critical component of living off the grid. Rainwater harvesting systems can collect and store rainwater for household use, while greywater systems recycle water from sinks, showers, and washing machines for use in irrigation. For those in suitable locations, wells can provide a steady supply of fresh water, though it's important to consider purification methods to ensure water safety.

Waste management on an off-grid homestead involves composting organic waste and recycling as much as possible. Composting reduces the amount of waste sent to landfills and produces valuable compost that can enrich the soil in your garden. Some off-grid homesteaders also explore more advanced waste management systems, such as biogas digesters, which can convert organic waste into energy.

Food production is the most rewarding aspect of living off the grid. Homesteaders can produce a significant portion of their food by cultivating a vegetable garden, raising livestock, and perhaps keeping bees. This reduces the carbon footprint associated with food transportation and ensures that the food is fresh, nutritious, and free of harmful chemicals. Moreover, preserving food through canning, drying, and fermenting can ensure a steady food supply year-round.

Living off the grid on a backyard homestead has its challenges. It requires a significant investment of time, resources, and energy. However, the rewards of a sustainable, self-sufficient lifestyle are immeasurable. It offers a profound sense of connection to the natural world, a deep appreciation for the resources we often take for granted, and the satisfaction of knowing that one is living in a way that is kind to the planet.

Chapter Summary

- Sustainable living on a backyard homestead emphasizes reducing waste through reducing, reusing, recycling, and composting, transforming our perspective on resources.
- Conscious purchasing decisions, such as choosing items with minimal packaging and made from sustainable materials, are key to reducing waste.
- Reusing items, such as repurposing glass jars for storage or old T-shirts for rags, fosters a culture of resourcefulness and innovation.
- Effective recycling requires understanding local programs and prioritizing easily recyclable materials to avoid contaminating the recycling stream.

- Composting kitchen scraps and yard waste closes the loop by returning nutrients to the soil, enhancing garden health, and reducing landfill waste.
- Eco-friendly homestead products, including tools made from sustainable materials and natural pest control methods, support a healthier ecosystem.
- Sustainable energy practices, like installing solar panels and wind turbines, reduce carbon footprints and can offer long-term savings and environmental benefits.
- Community involvement and education amplify the impact of sustainable living efforts, fostering a culture of sustainability and self-reliance through workshops, local markets, and community gardens.

12

HOMESTEAD PLANNING AND MANAGEMENT

A sustainable farm with solar panels and wind turbines.

Year-Round Planning

Establishing a backyard homestead requires a passion for self-sufficiency and a strategic approach to planning and management that spans the entire year. Year-round planning is the backbone of a successful homestead, ensuring every season brings its own set of tasks, goals, and rewards. This approach allows homesteaders to maximize their resources, manage their time effectively, and achieve a sustainable lifestyle.

The essence of year-round planning lies in understanding the cyclical nature of farming and homesteading activities. Each season has unique demands, from planting and harvesting to maintenance and preparation for the coming months. Spring is often the busiest season, filled with planting and early harvesting. It's a time to start seedlings, prepare garden beds, and set the foundation for the year's productivity. Summer follows its peak growth and harvest periods, requiring diligent care, watering, and pest management to ensure the health of crops and livestock. Fall brings another harvest season, alongside the need to preserve the bounty and prepare the homestead for the colder months. Winter, while seemingly quieter, is crucial for planning the following year's crops, repairing equipment, and focusing on indoor projects.

Effective year-round planning also involves understanding your local climate and ecosystem. For instance, knowing the first and last frost dates is vital for planting schedules. Similarly, being aware of local wildlife and their habits can help you plan defenses for your crops and livestock. Additionally, understanding the natural resources available on your land can guide decisions on water management, soil conservation, and sustainable energy sources.

To implement a successful year-round plan, homesteaders should start with a comprehensive calendar outlining the necessary tasks for each season. This calendar should include agricultural activities, infrastructure maintenance, financial planning, and personal learning goals. Regularly reviewing and adjusting this plan is essential, as it allows for incorporating new knowledge, adaptation to unexpected challenges, and optimizing processes.

Moreover, year-round planning is not just about the physical aspects of homesteading; it also encompasses the financial and emotional well-being of the homesteader. Balancing the demands of the homestead with personal health, family time, and community involvement is crucial for long-term sustainability. This holistic approach ensures that the homestead is not just a place of work but also a source of joy, fulfillment, and connection to the natural world.

In conclusion, year-round planning is dynamic and integral to managing a backyard homestead. It requires a deep understanding of the land, a commitment to sustainable practices, and a willingness to adapt and learn. By embracing this approach, homesteaders can achieve their goals of self-sufficiency and sustainability and enrich their lives and the environment around them. As we move forward, the next logical step in ensuring the success of our homestead is to delve into the intricacies of budgeting and managing expenses, ensuring that our endeavors are not only environmentally sustainable but financially viable.

Budgeting and Expenses

Transitioning from strategic considerations to practical budgeting and expenses is essential for effective homestead management. The financial health of your backyard homestead influences daily operations and future project feasibility, so managing your budget is crucial.

Start by creating a clear budget, listing all income sources and anticipated expenses. Be realistic in your estimates to avoid financial strain. This involves considering income from produce, eggs, honey, handmade goods, and expenses like seeds, feed, and maintenance.

Keeping a detailed record of expenses is vital for staying within budget and identifying cost-reduction areas. Use a spreadsheet or budgeting software designed for small-scale farming. Update this record regularly to reflect expenses and income, adjusting your budget as needed.

Prioritize investments that offer long-term benefits, such as quality tools and perennial plants. Over time, these can be more cost-effective than cheaper alternatives or annuals that need replanting each season. This approach ensures sustainability and efficiency.

If considering debt, proceed with caution. Loans or credit can fund significant investments like greenhouse construction but ensure the return justifies the borrowing cost. This strategy should enhance productivity and sustainability.

Building a reserve fund is critical for managing unexpected expenses, such as emergency repairs. This financial cushion ensures unforeseen costs don't compromise operations. It's an essential part of financial planning.

Adopt cost-effective practices like composting, rainwater harvesting, and seed saving. DIY projects can save money, provided you have the skills and resources. These practices reduce expenses and promote sustainability.

Consider diversifying your income streams to increase revenue potential and provide a buffer against failure. Exploring different markets, offering workshops, or venturing into online sales can enhance financial stability. Diversification is critical to resilience.

In conclusion, disciplined and strategic financial planning is foundational to the success and sustainability of your backyard homestead. As we discuss time management and efficiency, remember the close tie between financial health and effective time management. This approach ensures the growth and resilience of your homestead for years to come.

Time Management and Efficiency

In backyard homesteading, mastering the art of time management and efficiency is akin to discovering the secret garden of productivity. It's about making the most of the daylight hours, ensuring that every task, no matter how small, contributes to the overarching goal of self-sufficiency. This section delves into strategies and practices to transform your homestead from a demanding taskmaster into a well-oiled machine of productivity and satisfaction.

Firstly, understanding the rhythm of your land and its cycles is paramount. Nature operates on its timetable, and successful homesteaders align their activities with these natural cycles. This means planting and harvesting by the seasons, caring for animals based on their natural behaviors and needs, and even performing maintenance tasks when the weather and season dictate their necessity. By syncing your activities with these natural rhythms, you work more efficiently and increase the yield and health of your plants and animals.

Creating a prioritized task list is another cornerstone of effective time management. Not all tasks are created equal, and it is essential to recognize which tasks are critical and which can wait. Prioritization should be based on factors such as seasonality, the needs of plants and animals, and the sustainability of your homestead. For instance, planting or transplanting seedlings may take precedence in the spring, while harvesting and preservation dominate the autumn months.

Moreover, the implementation of systems and routines significantly enhances efficiency on a homestead. Systems can range from composting setups that reduce waste and nourish your soil to rainwater collection systems that ensure your plants are watered even in dry spells. Routines, such as regular feeding times for animals and daily garden inspections, help in the early detection of issues before they escalate, saving time and resources in the long run.

Another aspect often overlooked is the power of delegation and community. Homesteading is not a solitary journey. Involving family members in daily tasks lightens the load and instills a sense of responsibility and connection to the land. Furthermore, engaging with the wider homesteading community through cooperative purchasing, skill swaps, or collective labor efforts can significantly enhance efficiency and productivity.

Lastly, embracing technology and innovation can lead to substantial time savings and increased efficiency. Technology can be a powerful ally in managing a homestead, from drip irrigation systems that save water and time to apps that help track planting schedules and livestock health. However, it's essential to balance the benefits of technology with the principles of sustainability and self-sufficiency that lie at the heart of homesteading.

In conclusion, time management and efficiency in backyard homesteading are not about rushing through tasks or cutting corners. Instead, they are about thoughtful planning, understanding the natural rhythms of your environment, and leveraging both community and technology to work smarter, not harder. By adopting these practices, you can transform your homestead into a model of productivity and a source of deep personal satisfaction.

Record Keeping and Documentation

In transforming your backyard into a thriving homestead, meticulous record-keeping and documentation are indispensable tools. Often overlooked in the initial excitement of planting seeds and raising livestock, this practice is the backbone of efficient homestead management. It ensures that your efforts are fruitful but also sustainable and scalable.

At its core, record-keeping involves tracking various aspects of your homestead operations. This includes, but is not limited to, planting schedules, harvest yields, livestock health records, equipment maintenance logs, and financial expenditures and income. The purpose of maintaining such records is twofold. Firstly, it provides a clear snapshot of your homestead's current status, allowing for informed decision-making. Secondly, it offers invaluable insights into trends and patterns over time, which can inform future planning and improvements.

Starting with the garden, a simple journal can be transformative. Documenting what you plant, when, and where, alongside notes on weather conditions, pest issues, and harvest dates, can help refine your planting strategy year after year. This historical data becomes a guide, helping you to understand which crops thrive in your specific conditions and how to rotate them to maintain soil health.

For those raising livestock, health and productivity records are crucial. Tracking vaccinations, feed types and amounts, breeding cycles, and any health issues ensures your animals' well-being and impacts the quality and quantity of the produce they provide. Over time, these records can highlight patterns in health issues or productivity dips, enabling preemptive measures rather than reactive ones.

Financial documentation, though perhaps less appealing, is equally vital. Keeping detailed accounts of all income generated from your homestead, be it from selling produce, eggs, or handmade goods, as well as all expenses, offers a clear picture of your homestead's economic health. This financial clarity is essential for sustainability, allowing you to identify profitable ventures, cut unnecessary costs, and plan for future investments.

In today's digital age, numerous tools and software are available to simplify this process. From mobile apps designed for garden planning to comprehensive farm management software, technology can streamline record keeping, making it less daunting and more efficient. However, the charm and simplicity of a handwritten journal, with its sketches and personal notes, still hold value for many homesteaders.

As your homestead grows, these records become the foundation for scaling up. They allow you to assess the feasibility of expanding certain areas of your homestead, introduce new ventures with a clear understanding of the required resources and potential challenges, and ensure that growth is managed sustainably without compromising your land's or livestock's health.

In essence, diligent record-keeping and documentation are not mere administrative tasks but strategic tools that empower you to nurture a thriving, sustainable homestead. By embracing this practice, you lay the groundwork for meeting your current goals and expanding your vision for the future.

Scaling Up Your Homestead

As your backyard homestead begins to flourish, you may contemplate the following steps to expand your operations. Scaling up your homestead is exciting, but it requires careful planning and management to ensure success. This section will guide you through the considerations and strategies for effectively scaling up your backyard homestead.

Firstly, assess your current capacity and resources. Understand the limitations and potentials of your space, time, and budget. Expansion doesn't always mean acquiring more land; it can also mean optimizing your current space through vertical gardening, companion planting, or integrating more efficient systems like drip irrigation.

Next, set clear goals for scaling up. Are you looking to increase your food production, diversify the types of crops and livestock, or start a small-scale agricultural business? Specific objectives will help you plan the steps and resources required to achieve them.

Financial planning is crucial when scaling up. Consider the initial investment needed for additional resources, such as seeds, tools, livestock, or infrastructure improvements. It's also wise to project the potential return on investment, especially if you plan to sell your produce or products. Creating a detailed budget will help you manage expenses and assess the financial viability of your expansion.

Expanding your homestead may also require additional knowledge and skills. Whether learning about new crop varieties, animal care, or sustainable farming practices, investing in your education will pay dividends. Consider attending workshops, joining local farming groups, or seeking mentorship from experienced homesteaders.

As you plan your expansion, think about the long-term sustainability of your homestead. Incorporate practices that enhance soil health, conserve water, and promote biodiversity. Scaling up offers an opportunity to deepen your commitment to sustainable and regenerative agriculture.

Finally, be prepared to adapt. As you implement your plans, you may encounter unexpected challenges or opportunities. Stay flexible and open to adjusting your strategies as needed. Remember, scaling up is a journey that requires patience, perseverance, and a willingness to learn and grow.

By thoughtfully planning and managing the scaling up of your backyard homestead, you can increase your self-sufficiency, contribute to your community, and enjoy the rewards of a thriving homestead. Remember that expansion should enhance your homesteading experience, not overwhelm it. With careful consideration and strategic planning, you can successfully scale up your backyard homestead and achieve your goals.

Dealing with Setbacks

Expanding and managing your backyard homestead will require setbacks. These challenges are inevitable, but you can overcome them with resilience and the right strategies. This section will guide you through common issues and provide practical advice to keep your homestead thriving.

Weather can be unpredictable, affecting your plants and livestock in various adverse ways. Droughts, floods, unexpected frosts, and heatwaves can all cause significant damage.

Implement water-saving techniques like rainwater harvesting and drip irrigation to combat these issues. Using row covers and shade cloth and ensuring proper ventilation for animal housing can protect against frost and heat, and having an emergency plan for extreme weather is crucial.

Pests and diseases can quickly turn a healthy garden or flock into a struggling one. Employing Integrated Pest Management (IPM) strategies, which include biological controls, natural predators, and organic pesticides, can help manage these issues in an eco-friendly way. Regular health checks and quarantining new additions to your homestead can prevent outbreaks from taking hold.

As your homestead grows, so does the complexity of its management, potentially leading to operational overwhelm. Creating a robust system for tracking tasks, expenses, and yields can help manage this complexity. Automating repetitive tasks and delegating responsibilities can reduce pressure, allowing you to focus on strategic improvements.

Financial setbacks, such as unexpected expenses or market downturns, can impact your homestead's stability. Building a financial buffer, diversifying income streams, and applying lean management principles provide some security. Seeking out grants, loans, and subsidies, as well as maintaining meticulous financial records, can also aid in managing your budget effectively.

Change is constant in homesteading, whether due to evolving consumer preferences, new regulations, or climate shifts. Staying informed through continuous learning and networking with other homesteaders is vital. Being willing to adapt your strategies, whether by trying new crops or exploring alternative markets, can keep you ahead of these changes.

In conclusion, setbacks are a natural part of homesteading. They challenge you to be resilient and innovative and ultimately contribute to your success. By preparing for potential challenges and responding proactively, you can navigate these hurdles and maintain a sustainable and fulfilling backyard homestead.

Chapter Summary

- Year-round planning is essential for a successful backyard homestead, understanding the cyclical nature of farming activities and adjusting tasks seasonally.
- Effective planning requires knowledge of the local climate, ecosystem, and available natural resources to optimize planting schedules and manage resources sustainably.
- A comprehensive calendar outlining agricultural activities, infrastructure maintenance, and personal goals is crucial, with regular reviews to adapt and optimize processes.
- Financial and emotional well-being, along with balancing homestead demands with personal health and community involvement, are key to sustainable homesteading.
- Budgeting and expense management are foundational, involving realistic income and expense estimates, prioritizing long-term investments, and maintaining a reserve fund for unforeseen costs.
- Time management and efficiency in homesteading rely on understanding land rhythms, prioritizing

tasks, implementing systems and routines, and leveraging community and technology.

- Meticulous record-keeping and documentation of homestead operations are indispensable for informed decision-making and future planning.
- Scaling up a homestead requires assessing current resources, setting clear goals, financial planning, acquiring new knowledge, and adapting to challenges and opportunities.

THE FUTURE OF BACKYARD HOMESTEADING

Peaceful countryside with a greenhouse and chickens.

Reflecting on the Journey

As we stand at the threshold of the future, looking back on the journey of backyard homesteading, it's essential to pause and reflect on the path we've traversed. This journey, marked by the sweat of our brows and the dirt under our fingernails, has been more than just about cultivating land; it's been about cultivating life.

Backyard homesteading has transformed from a mere hobby or trend into a lifestyle that resonates with the core of sustainable living. It has taught us the value of self-reliance in an era where convenience often trumps sustainability. We've learned to appreciate the cycles of nature, understand the rhythm of the seasons, and respect the balance of life that thrives in our backyards.

This journey has not been without its challenges. There were days when the soil seemed too stubborn to till and seasons when the harvests were meager. Yet, these obstacles only strengthened our resolve and deepened our connection to the land. They reminded us that homesteading is not just about the yield but the process and the profound lessons learned along the way.

The future of backyard homesteading looks promising, with more individuals and families awakening to the joys and benefits of this fulfilling practice. Technology and innovation continue to open new avenues for sustainable living, making it more accessible and efficient. However, the essence of homesteading remains rooted in the timeless principles of hard work, patience, and stewardship of the earth.

As we move forward, let us carry the lessons of the past with us. Let us continue to share our knowledge and experiences with the community, fostering a spirit of cooperation and mutual support. The impact of our homesteading journey extends beyond the confines of our backyards, influencing our families, communities, and the environment in profound ways.

In embracing the future of backyard homesteading, we are

not just cultivating our gardens but nurturing a legacy of sustainability and resilience for generations to come. This journey, with its highs and lows, has been a testament to the human spirit's capacity for growth and adaptation. It is a journey worth celebrating, a journey worth continuing.

The Impact of Homesteading on Family and Community

As we delve into the essence of backyard homesteading, it becomes evident that its impact stretches far beyond the confines of one's property, weaving itself into the fabric of family life and community spirit. The journey of transforming a simple backyard into a thriving homestead is not just about cultivating the land; it's about nurturing relationships and building a sense of belonging and responsibility towards one another and the environment.

For families, the homesteading lifestyle offers a unique platform for learning and growth. It fosters a culture of self-reliance and problem-solving, where each family member can contribute their skills and learn new ones. Children raised in this environment often develop a profound understanding of the cycles of nature, the value of hard work, and the importance of sustainability. They witness firsthand the fruits of their labor, from planting seeds to harvesting crops, which instills a sense of pride and accomplishment in them. This hands-on approach to learning encourages curiosity and a deep appreciation for the natural world.

Moreover, backyard homesteading has the potential to strengthen family bonds. Working together towards a common goal creates opportunities for quality time and shared experiences. Families can connect on a deeper level in these moments, whether tending to the garden, caring for animals, or preserving the harvest. The challenges and triumphs experienced along the way become cherished memories and valuable life lessons passed down through generations.

The influence of backyard homesteading extends into the broader community as well. It can catalyze neighborhood collaboration and support. Homesteaders often share surplus produce, seeds, and plants with their neighbors, fostering a culture of generosity and mutual aid. This exchange helps build resilient local food systems, encourages social interaction, and strengthens community ties. Additionally, homesteading initiatives can inspire community projects, such as communal gardens, farmers' markets, and educational workshops, further enriching the area's social fabric.

Another significant contribution of backyard homesteading to the community is the revival of traditional skills and knowledge. Once commonplace, skills such as gardening, preserving food, and carpentry have faded in the modern era. Homesteading revives these practices, preserving them for future generations while promoting a more sustainable and self-sufficient lifestyle.

In essence, the impact of backyard homesteading on family and community is profound and multifaceted. It nurtures a sense of responsibility, fosters connections, and promotes a sustainable way of living that benefits the individual homesteader and the community as a whole. As we look toward the future of backyard homesteading, it's clear that its potential to transform lives and communities is immense. The journey of homesteading, with all its challenges and rewards, is a testament to the resilience and creativity of the human spirit, offering a path towards a more connected, sustainable, and fulfilling way of life.

Challenges and Rewards

The path of a backyard homesteader is only sometimes smooth; it is paved with obstacles that test resilience, adaptability, and commitment. Yet, these very challenges make the rewards even more gratifying.

One of the primary challenges faced by homesteaders is the steep learning curve. Whether mastering the art of vegetable gardening, understanding the nuances of animal husbandry, or learning how to preserve food, each skill requires time, patience, and a willingness to learn from mistakes. Nature's unpredictability adds another layer of complexity, with weather conditions, pests, and diseases posing constant threats to crops and livestock. Moreover, the initial financial investment and ongoing maintenance costs can be significant, requiring careful planning and budgeting.

Despite these hurdles, the rewards of backyard homesteading are immense and multifaceted. A profound sense of accomplishment comes from growing your own food, reducing dependence on commercial food systems, and living more sustainably. This lifestyle fosters a deep connection with the natural world, encouraging a rhythm of life that is more in tune with the seasons and cycles of the earth. The health benefits are also noteworthy, with access to fresh, organic produce and the physical activity involved in gardening and animal care contributing to overall well-being.

Furthermore, backyard homesteading offers invaluable self-reliance, problem-solving, and creativity lessons. It cultivates a spirit of innovation, as homesteaders often find themselves devising unique solutions to challenges, whether repurposing materials for garden beds or developing efficient systems for water conservation. Another significant reward is the sense of community that emerges within the family and with fellow homesteaders. Sharing knowledge, experiences, and the fruits of one's labor fosters a sense of belonging and mutual support that is increasingly rare in today's fast-paced world.

As we look to the future, it's clear that the challenges and rewards of backyard homesteading will continue to evolve. With technological advancements, new opportunities for efficiency and sustainability are emerging, promising to shape the next chapter of this enduring lifestyle. The journey of a backyard homesteader is one of constant learning and adaptation. Still, it is also a testament to the enduring human spirit's desire to connect with the earth and live in harmony with nature.

Advancements in Homesteading Techniques

Technology has become a surprising ally in the quest for sustainability and efficiency in recent years. Innovative gardening tools, for instance, have revolutionized how we approach the cultivation of our land. Soil sensors can now provide real-time feedback on moisture levels, pH balance, and nutrient content, allowing for precise adjustments to ensure optimal plant growth. This technology, once the purview of commercial agriculture, is becoming increasingly accessible to the everyday homesteader, enabling us to produce more with less effort and fewer resources.

Another significant advancement is in the realm of water

conservation and management. Rainwater harvesting systems have evolved from simple barrels to sophisticated systems that can collect, filter, and store water for irrigation, reducing our reliance on municipal supplies or well water. Coupled with drip irrigation technology, which delivers water directly to the plant roots with minimal waste, these systems exemplify how modern innovations support sustainable homesteading practices.

Renewable energy sources have also found their place in the backyard homestead, further reducing our carbon footprint and enhancing self-sufficiency. Solar panels can power everything from water pumps to lighting; small-scale wind turbines can supplement energy needs. These renewable energy systems are becoming more affordable and user-friendly, making it feasible for homesteaders to harness the power of the elements.

Integrating aquaponics and hydroponics into backyard homesteading presents a leap forward in food production. These systems, which grow plants without soil, offer a sustainable alternative to traditional gardening, using significantly less water and space. Fish waste provides a natural nutrient source for the plants in aquaponics systems, creating a closed-loop ecosystem that mimics nature's efficiency. These systems yield a diverse array of produce and open the door to year-round gardening, irrespective of climate.

As we look to the future, it's clear that the potential for backyard homesteading is boundless. The advancements in techniques and technology not only make homesteading more accessible but also more impactful. By embracing these innovations, we can enhance our self-reliance, reduce our environmental impact, and cultivate a deeper connection with the land.

In embracing these advancements, we also pave the way for

a legacy of sustainability. The knowledge and practices we develop and refine today will be the inheritance of future generations of homesteaders. As we continue to innovate and adapt, we ensure that the art of homesteading not only endures but thrives, fostering a resilient and sustainable relationship with our environment for years to come.

Sustaining Your Homestead for Future Generations

The essence of homesteading is not just in self-sufficiency but in creating a legacy that outlives our efforts, ensuring that the knowledge and skills we cultivate are passed down and built upon.

Sustaining your homestead for future generations begins with a commitment to eco-friendly practices. This involves adopting methods that replenish the resources we use rather than depleting them. Composting, rainwater harvesting, and renewable energy sources are just a few ways to minimize your homestead's environmental impact. By integrating these practices, you make your homestead more sustainable and teach the next generation the importance of living in harmony with nature.

Another critical aspect is the preservation of heirloom seeds and the cultivation of biodiversity. By choosing to plant heirloom varieties, you're not just growing food; you're preserving plant genetics passed down through generations. This biodiversity is crucial for resilience against pests, diseases, and changing climate conditions. Encouraging local wildlife to thrive by creating habitats and food sources also contributes to a balanced ecosystem on your homestead.

Education plays a pivotal role in sustaining your homestead

for future generations. This doesn't necessarily mean formal education; instead, it means sharing knowledge and experiences with family, friends, and the community. Workshops, community gardens, and social media are powerful tools for spreading the wisdom of homesteading practices. By fostering a community of like-minded individuals, you ensure that the skills and values of homesteading are preserved and evolved.

Finally, it's essential to embrace innovation and adaptability. As we've seen in the advancements of homesteading techniques, staying open to new ideas and technologies can significantly enhance your homestead's efficiency and sustainability. Whether it's through adopting new gardening methods, water-saving technologies, or sustainable building materials, being adaptable ensures that your homestead can meet future challenges.

In conclusion, sustaining your homestead for future generations is a multifaceted endeavor that requires a commitment to environmental stewardship, education, community building, and adaptability. By embedding these principles into the fabric of your homesteading practices, you lay the groundwork for a legacy that nurtures the land and the community for future generations. As we move forward, let us carry with us the responsibility to reap the earth's benefits and enrich it, ensuring a prosperous and sustainable future for all who follow in our footsteps.

Final Thoughts and Encouragement

As we stand at the threshold of the future, looking back at the journey of backyard homesteading, it's clear that what we've embarked upon is not just a trend but a transformative lifestyle that reconnects us with the earth and our food sources. The path you've chosen or are considering is one of resilience, sustainability, and profound satisfaction. It's about taking control of your food supply, reducing your carbon footprint, and nurturing a deeper connection with nature.

The journey of a backyard homesteader is filled with trials and triumphs. There will be days when your garden's bounty exceeds all expectations and others when unforeseen challenges test your resolve. Remember, every setback is a learning opportunity, and every success, no matter how small, is a step towards self-sufficiency and environmental stewardship.

As you move forward, let your homestead evolve with you. Adapt and innovate as you learn more about sustainable practices and the unique needs of your land. The beauty of backyard homesteading is that it's not a one-size-fits-all model but a canvas for creativity and personal expression.

Encourage others to start their journey into homesteading by sharing your experiences and the benefits you've reaped. Whether through community workshops, social media, or casual conversations, your story can inspire and empower others to take steps towards a more sustainable lifestyle.

Lastly, remember to enjoy the journey. Take time to appreciate the simple pleasures of homesteading—the morning dew in your garden, the taste of freshly harvested produce, and the joy of sharing your bounty with loved ones. These moments are the true essence of backyard homesteading.

As we look to the future, it's clear that the principles of backyard homesteading—sustainability, self-sufficiency, and community—will play a crucial role in shaping a more resilient and environmentally conscious society. Your commitment to this lifestyle is a beacon of hope and a testament to the positive change individuals can create in their corner of the world.

So, as you close this book and step back into your garden, remember that you are part of a growing movement towards a sustainable future. Your backyard homestead is not just a piece of land but a living, breathing testament to the change we wish to see. Keep nurturing it, keep growing, and keep inspiring those around you. The future of backyard homesteading is bright and in your hands.

Your Feedback Matters

Thank you for joining me on this journey. If the book inspired you, please share your thoughts by leaving a review on Amazon using the QR code below. Your feedback is invaluable and helps guide others. I'm grateful for your time and hope the insights you've gained enrich your quest for knowledge.

SCAN ME

ABOUT THE AUTHOR

Anthony Bennett is a passionate advocate for sustainable living and self-sufficiency. With a background in environmental science and years of hands-on experience in homesteading, his practical advice and innovative solutions for living off the land have inspired many to pursue a more sustainable and independent lifestyle. Through his "Self-Sufficient Living" series, including "How to Build the Perfect Backyard Homestead" and "How to Survive When the Grid Goes Down," Anthony shares his wealth of knowledge on creating resilient and productive homesteads. When he's not writing, Anthony tends to his homestead.